"Before AI Decides brings rare clarity to one of the defining challenges of our time. Stevens highlights the rapid deployment of AI and its remarkable abilities, insisting that vigilant human attention remain essential to guard against people being overridden as intelligent systems shape decisions that matter."

ERIK VIIRRE, MD, PHD
DIRECTOR, CENTER FOR HUMAN IMAGINATION, UCSD
https://imagination.ucsd.edu/erik-viirre/

"Before AI Decides is a clear-eyed wake-up call about protecting human judgment in an automated world. Stevens shows how AI shifts control not through takeover, but through subtle, everyday decisions. He is no AI doomster. Acknowledging AI's promise, he argues for guardrails that keep people in the loop. His Declaration of Human Rights for the AI era and Nine Principles offer a practical framework for navigating AI with awareness and care."

ANNE LEER
AI EXPERT ON POLICY & GOVERNANCE
www.linkedin.com/in/anneleer

"Having given a lifetime helping humanity, acting with humility, wisdom, and integrity with big picture ideas, Payson Stevens' seminal work on conscientiously incorporating AI into our lives is a gift beyond measure."

KEN DRUCK, PHD
BEST-SELLING AUTHOR
www.kendruck.com

"Human progress has always involved risk and reward. AI now confronts us with something new: autonomy at scale without visibility or reversal. That combination may mark a point of no return. Before AI Decides makes this danger clear and urgent."

LEE STEIN
ENTREPRENEUR & INVENTOR
https://stein.to

"AI's current trajectory threatens something fundamental: the human capacity to ask 'why.' This book offers a clear, grounded, and practical way to keep human intelligence alive and amplified in our encounters with AI."

JESSICA SEDDON, PH.D.
SENIOR LECTURER, YALE JACKSON SCHOOL OF GLOBAL AFFAIRS
https://jscaseddon.co

A clear-eyed guide for anyone trying to stay human in a world increasingly shaped and dominated by AI. Stevens lays out an essential framework for safeguarding our rights, our humanity, and our sanity, as this new technological juggernaut unfolds."

CHARLES M. LEVINE
FORMER VICE-PRESIDENT, RANDOM HOUSE
https://www.linkedin.com/in/charleslevine/

BEFORE AI DECIDES

Nine Ways to Stay Human

PAYSON R. STEVENS

TARANG PRESS

Tarang Press
P.O. Box 810
Del Mar, CA 92014
tarangpress.com

Cover: Charanjit Singh - corepr.com
Book Design: J.C. Griffith - jcg.studio

www.paysonrstevens.com
payson.stevens@gmail.com

ISBN: 9798994760307

To the memory of those who shaped my mind to think critically and ask the right questions.

"In every deliberation, we must consider the impact of our decisions on the next seven generations."

GREAT LAW OF PEACE
HAUDENOSAUNEE/IROQUOIS CONFEDERACY

The Seven Generations principle, rooted in Haudenosaunee oral law, was first recorded in English by scholars and Indigenous historians in the late nineteenth and early twentieth centuries, as the Great Law of Peace was translated from countless generations of Native American oral tradition into written form.

"An unexamined life is not worth living."

SOCRATES
ANCIENT GREEK PHILOSOPHER,
CLASSICAL ATHENS. 470 BCE TO 399 BCE

TABLE OF CONTENTS

How This Book is Organized

This book is designed to be read in layers.

The Nine Pillars form the core of the book. The storys, interludes, and reflections around them provide context, consequence, and meaning.

The book subtitle calls them "Nine Ways" for a simple reason: each Pillar is meant to be usable. In these pages, the Pillars are the names of the principles and a foundation to help you navigate AI in your life. Those same principles show up as choices you can notice and practice, especially when AI feels effortless or final. The goal is not to memorize a framework. It is to stay oriented and to keep human judgment active as AI enters ordinary life.

You can read straight through or return to sections as needed.

The goal is not mastery. It is orientation and personal power.

PART I: STORIES FROM A CHANGING WORLD
How invisible technologies reshape everyday life

PART II: THE NINE PILLARS FOR STAYING HUMAN
A practical framework for life inside AI systems

PART V: REFLECTIONS

What this moment asks of us as humans

APPENDICES

What this moment asks of us as humans

FOREWORD

Why This Book Is an Urgent Wake-Up Call

It no longer matters who you are or where you live. Rapid advances in artificial intelligence are already changing how we live, learn, heal, work, and play. **The AI genie is out of the bottle.** The question is no longer whether AI will shape our future, but who will be the masters when it does.

In *Before AI Decides*, Payson R. Stevens makes a clear and necessary argument: authority is quietly shifting from people to systems, often without consent, without explanation, and without meaningful opportunity for appeal. His call to become "Directors of AI" is not rhetorical. It is a practical framework for reclaiming responsibility before judgment slips out of human hands.

Most of us experience this transition as convenience. A recommendation appears. A form fills itself. A decision arrives fully formed, carrying the weight of authority but offering no explanation. These moments feel ordinary, even helpful. They are anything but. They mark the point at which assistance begins to turn into control.

I have spent my career deeply engaged in the digital transformation of industry, government, education, and culture, watching technological systems reshape our world. I have seen innovations promise liberation and deliver control. I have witnessed convenience become dependency, and efficiency replace judgment.

What Stevens identifies with unusual clarity is the pattern beneath these shifts: decision-making authority migrates first invisibly, then irreversibly. He names this asymmetry clearly. Systems now operate faster and at greater scale than people can see, question, or correct. This is not a future risk. It is already shaping who gets hired, who receives care, who qualifies for opportunity, and whose voice is heard.

Stevens gives this imbalance a name: **Shadow Hydra**, a system-level asymmetry in which decisions are made faster and at greater scale than people can meaningfully observe, question, or correct. What makes it dangerous is not intent, but momentum.

What distinguishes this book is that it refuses to treat AI as merely a technical problem. Payson understands that AI alters the relationship between humans and judgment itself. When systems are trusted because they work, authority moves quietly. When decisions arrive faster than explanation, consent erodes. When a doctor defers to "what the system recommends," when a navigation app becomes the unquestioned authority during a wildfire, when students are tracked before anyone asks if tracking serves them—these are not efficiency gains. They are transfers of power that happen without consent, often without awareness.

Payson brings rare perspective to this work. He has witnessed multiple technological revolutions, communicated complex science to public audiences, and tracked climate disruption long enough to recognize how systems tip quietly before consequences become visible. He has lived and worked in remote Himalayan valleys, studying how traditional cultures absorb technological change.

He writes with the clarity that comes from having watched multiple technological eras reshape society, and from understanding that this one is different.

This is why the **Nine Pillars** at the heart of this book matter. They are not abstractions or policy slogans. They are practical anchors for preserving human agency in systems increasingly optimized for speed rather than understanding. Seek the Truth. Show the Work. Let Humans Decide. These are not nostalgic values. They are survival skills for democratic societies.

Importantly, this book is not anti-technology. Stevens is no AI doomster. He is clear about AI's potential to serve humanity, and he is transparent about his own use of AI in writing this book. He models what he argues for: humans directing tools, not deferring to them. This transparency matters. As AI becomes embedded in creative and intellectual work, pretending it is absent helps no one. What matters is whether judgment remains human, and whether governance keeps pace with capability.

The deepest challenge Stevens identifies is democratic. When decisions move into code faster than public debate can follow, democracy does not collapse. It fades. Accountability thins. Consent becomes impossible. This is not a failure of innovation. It is a failure of governance.

At a moment marking 250 years of American democracy, institutions designed for human judgment now confront machine-speed automation. This collision will define whether democratic values persist or whether power quietly concentrates in systems beyond public reach. This is not a choice between progress and stagnation, but between accountable innovation and unaccountable power.

Before AI Decides arrives at the right moment because it insists that the future remains open. Not inevitable. Not predetermined. Open. The practices Stevens offers are ways of staying present when systems encourage passivity. His *Declaration of Human Rights in the Age of AI* provides a moral frame for what must not be automated away.

Staying human is no longer passive. It is an active choice. This book helps readers notice when authority shifts, slow decisions that feel inevitable, and reclaim responsibility while it still exists.

Read this book not as prophecy, but as orientation. Let it help you notice where judgment has moved, where agency still exists, and why the future remains open only by choice.

The intelligence that made us is still what will carry us forward. But only if we remember to use it.

Anne Leer
Global Policy and AI Infrastructure Expert
Chair, Oxford AI Leadership Forum
www.linkedin.com/in/anneleer

AUTHOR'S PREFACE

The Moment Everything Shifted

The doctor did not look at us when he spoke. He looked at his computer screen. My wife Kamla's health had been declining for months. We had seen specialists, run tests, and waited for calls that arrived slowly or not at all. Many appointments ended with more uncertainty and less clarity. That day, after a long pause, the doctor finally said what he believed would help.

"The system recommends seeing another specialist."

Not I recommend. Not we recommend. The system.

Something shifted in the room. Authority moved from a person to a process neither of us could see or question. The recommendation carried weight, but no explanation. It arrived fully formed, without anyone clearly responsible. Kamla was suffering. Time mattered. Yet an algorithm trained on millions of cases, many unlike hers, was influencing her care without context or accountability.

That moment stayed with me.

Around the same time, I was talking with my friend Lee Stein, a careful observer of what happens when change accelerates faster than people and institutions can absorb it. We were discussing AI not as a future possibility, but as something already shaping everyday life through rankings, recommendations, summaries, and defaults. Lee asked a question that became central to this book: "What can a book possibly do when everything is changing this fast?

The answer was not to predict the future. It was to help people stay oriented inside it.

From Attention to Decisions

We are living through a period of acceleration unlike anything I have witnessed, and perhaps unlike anything in human history. AI now influences what people see, what they qualify for, how they are judged, and which opportunities reach them. Much of this happens quietly. People feel they are choosing freely, even as systems narrow the path ahead.

AI is advancing faster than laws, institutions, and cultural norms can adapt. Companies are racing to dominate it, much as they once raced to dominate social media. But this time is different.

Social media reshaped attention, with consequences we are still struggling to contain. It rewarded outrage, comparison, and speed. It trained people to react before reflecting. The damage was not sudden. It accumulated quietly, reshaping habits, expectations, and mental health long before society understood what was happening. AI moves the pressure one layer deeper. It does not just shape what we notice. It increasingly shapes what is decided.

AI can support natural intelligence. It can also replace it gradually, without asking permission. The risk grows when people stop noticing where judgment has moved.

This book is about that shift. The defining challenge of the AI era is asymmetry: a small number of fast, automated systems shaping outcomes for millions of people who cannot see, question, or contest how decisions are made. As speed increases, reflection lags. Defaults harden into rules before anyone pauses to ask whether they should exist at all.

*In response, this book introduces the **Nine Pillars**, a human-scale framework distilled from my broader **Declaration of Human Rights in the Age of AI** (Appendix I). The Pillars are not technical rules or policy prescriptions. They are habits of attention, designed to keep judgment visible, contestable, and accountable as AI grows more powerful.*

*Later in the book, I name a pattern of risk that emerges when this imbalance goes unnoticed. I coined the term **Shadow Hydra** to describe a danger that multiplies gradually through delegation, convenience, and inattention, until control slips away without anyone ever deciding to give it up.*

This book is not about rejecting AI.

AI holds real promise. Used well, it can expand knowledge, reveal patterns we miss, support care, and help solve problems at scales no individual could manage alone. I use it daily, and it helped me write this book faster than any I have written before. But using AI intelligently is not about speed or convenience. It is about keeping human judgment present when it matters most.

A Life Watching Technology Change

At eighty, I have spent more than five decades working at the intersection of science, art, and technology. In the early 1980s, at the beginning of the digital revolution, my companies worked with scientists at NASA, NOAA, and the U.S. Geological Survey, helping translate complex Earth System Science into forms that could be understood beyond expert circles.

*That work led to pioneering multimedia projects, including **Arctic Data InterActive**, a CD-ROM science journal created long before such formats were common. My companies later published ten award-winning CD-ROM, educational titles that brought science and the environment to broad audiences. This work was recognized with a Presidential Design Award presented by President Bill Clinton in 1994.*

Having witnessed the birth of the digital age, the rise of the internet, and the steady reshaping of attention by screens, I believed I understood technological change. AI changed that assumption. AI is not just another tool. It operates at automated scale and acceleration. People do not. When systems move that fast, even small errors can multiply before anyone notices.

Human Intelligence Still Matters

Throughout this book, I use the terms human intelligence and natural intelligence to mean our evolved capacity for judgment shaped by experience, context, memory, empathy, and meaning. That intelligence is not outdated. It is essential. Speed can feel like understanding. It is not.

Kamla's health has been improving not because we surrendered to a pro-

cess, but because we used AI as a tool while keeping human judgment at the center. That experience clarified what is at stake.

The future is not decided all at once. It is shaped in small moments: in the settings we accept, in the choices we pause to question, and in habits we repeat until they feel normal. **Before AI Decides** is about staying awake in those moments. AI finds patterns. Our intelligence finds meaning.

Because when we stop owning decisions, AI begins governing us.

- Payson

WRITING WITH AI

This book argues that human judgment must remain central as AI grows more capable. That belief shaped not only what I wrote, but how it was written. I did not write this book alone. Claude, ChatGPT, and Gemini were part of the process. Not as authorities or voices to defer to, but as tools I worked with directly, prompted deliberately, and questioned constantly as the ideas took shape.

I used AI to test ideas, challenge assumptions, and explore alternate framings. It helped compress drafts, surface repetition, and expose where arguments drifted toward vagueness or false confidence. Sometimes it was genuinely useful. Other times, it produced confident mistakes or *hallucinations*: language that sounded plausible but was not grounded in fact or experience and had to be removed. Throughout the process, I remained responsible for every decision. That distinction matters.

AI often functioned much like a traditional ghostwriter. It helped research background material, draft early versions of composite stories, suggest language, reorganize thoughts, and accelerate revision. It did not invent my ideas or replace my voice. It helped ideas arrive faster. That metaphor made me uneasy, which is precisely why I chose to explain the process openly here.

Every story, argument, judgment, and conclusion in this book is mine. I decided what stayed and what went, checked sources, and questioned certainty. When AI output sounded persuasive without grounding in lived experience, I cut it. Every paragraph it touched was reviewed, revised, challenged, or rewritten multiple times. I kept returning to a simple question: Is this saying what I mean, or is it just saying something well?

Several terms and frameworks introduced in this book, including **Shadow Hydra** and **Become the Director of AI**, emerged through this process to name patterns I observed but could not adequately describe with existing language.

The sheer volume of output sometimes became its own hazard. Too many versions. Too many files. Too many promising directions competing at once. I told Kamla it felt like an *engaged stupor*: productive, busy, and strangely unmoored. Ideas arrived faster than judgment. Stopping, surfacing, and choosing again required deliberate effort.

That experience taught me something important. The real danger is not when AI is wrong. It is when it sounds right. Polished language can pass for understanding. Speed can replace thinking. Confidence can show up with no one responsible. Writing this book forced me to confront how easily authorship can blur. It was tempting to let AI speak for me when it sounded coherent and calm. Resisting that pull became part of the work. I had to slow down, reassert judgment, and make sure my voice, craft, and lived experience remained present. That dynamic mirrors the argument of this book.

AI excels at pattern recognition, synthesis, and scale. It is weak at meaning, responsibility, lived experience, and moral judgment. Used well, it can support human thinking. Used uncritically, it can invisibly replace it.

Throughout the book, I treated AI as something to engage with, not defer to. I asked it to explain itself and compared its outputs to real-world experience. It was a powerful tool that required oversight, not trust. In other words, I practiced what this book argues for. I learned to act as the director of AI, not its audience.

This book is the result of my judgment, reflection, revision, and choice. AI helped me move faster. It did not decide what this book means. Transparency matters. As AI becomes more present in creative, professional, and intellectual work, pretending it is not involved helps no one. The important question is not whether AI was used, but how.

Used openly, critically, and with human judgment intact, AI can be a powerful tool. Used invisibly or without question, it becomes a substitute for thinking.

This book calls for the first path.

LIVING WITH TECH THAT DECIDES FOR US

Most people do not choose AI. It is built into the technical framework they use every day.

Forms fill themselves out. Recommendations appear. Scores, rankings, and predictions shape what is offered, approved, delayed, or denied. These outcomes feel normal and efficient. Few people experience them as decisions being made. They feel like the way things work now. Nothing announces that judgment has moved.

This is how AI enters daily life. Not as a dramatic change, but as a persistent one. Tasks and choices once handled by people are absorbed into technologies designed to move fast, stay consistent, and operate at scale. At first, this feels helpful. Less effort. Less friction. Fewer steps to think through. Over time, something more important changes. Automated processes begin to shape which options are visible and which fade away. This book is about that change.

The Central Issue:
Asymmetry Before and After AI

Asymmetry is an imbalance in which decisions are made faster and at greater scale than people can see, question, or correct. Outcomes arrive as final. The reasons behind them remain distant or unreachable. **Imbalance is one of the core ideas of this book. It existed long before AI.**

» A bank changes its overdraft rules and fees
   appear before anyone can explain why.

» An airline cancels a flight and policy blocks refunds,
   even when no person caused the delay.

» A health insurer denies coverage based on fine
   print no one remembers agreeing to.

» A company restructures and jobs disappear
   through a spreadsheet, not a conversation.

» A school district changes testing standards and
   families learn only after scores drop.

In each case, the decision lands. The explanation does not. No algorithm is required for asymmetry to exist. What matters is how decisions are made, how quickly they travel, and how far removed they are from the people affected.

As these asymmetries compound, another recognizable pattern emerges. I have named it the **Shadow Hydra**. The term describes many separate systems making small, defensible decisions in isolation that combine into outcomes no one fully oversees or owns.

This pattern is dangerous because it hides risk in plain sight. No single decision appears reckless. No single tool looks out of control. Yet small distortions accumulate, and ownership dissolves across layers until no one can clearly explain, challenge, or reverse what has happened. What feels efficient at first can harden into a structure that resists accountability.

AI did not invent this imbalance. It accelerates it. The same pattern now appears in AI-mediated decisions:

» Credit is denied without explanation.

» Job applications vanish inside automated screening.

» Medical recommendations arrive stripped of context.

» Feeds quietly reshape what people see and come to believe.

Again, the decision feels complete, while the reasons remain partial, opaque, or unavailable.

AI can support human judgment. It can surface patterns, reduce overload, and improve care when used thoughtfully. But it can also replace judgment rather than assist it. Systems can standardize decisions that require context. They can scale errors as easily as insight. Temporary shortcuts can harden into permanent structures.

What changes is not just speed, but responsibility. The risk is not intelligence itself. The risk is losing our shared ability to decide what should happen and to know who is responsible when it does.

What You Will Get from This Book

This book helps you:

» **Notice when a system is deciding for you,**
   not just offering help

» **Question outcomes that arrive without explanation,**
   instead of accepting them as final

» **See where responsibility has shifted,**
   even when no one appears to be in charge

» **Protect judgment, fairness, and dignity**
   in everyday situations

» **Slow decisions that feel inevitable and reclaim**
   the ability to respond

The pages ahead are about keeping people's judgment present before decisions harden into structures.

WHY THIS BOOK NOW

Clarity Before Decisions Harden

You do not need another book about whether AI is good or bad. You need clarity before defaults lock in. AI advances every week. Systems grow more confident. Questioning starts to feel inefficient. Compliance starts to feel normal. The window for how AI affects your life is closing quietly.

This book is meant for that moment and gives you:

- » Six warning signs that systems are already deciding for you
- » Nine principles to protect human judgment
- » Three pressure points where delay becomes costly
- » Five habits you can use immediately

It approaches AI from two directions at once: lived experience and practical guidance. Both are present throughout. Read it for the stories, the frameworks, or the questions that help you notice what is changing. Use what fits your life, needs, and interests right now. **If you read nothing else, read the Nine Pillars.** Each takes about ten minutes. Together, they hold whether AI accelerates, stalls, or veers in unexpected directions.

This is not a prediction book.

It is an orientation book.

Most AI books ask what will happen.

This one asks who decides.

HOW TO USE THIS BOOK

Read What Matters. Skip What Doesn't

This book is designed to be used, not read straight through. You do not need technical knowledge. You do not need to read every page in order. You do not need to agree with everything here to benefit from it.

Some sections are reflective. They slow the pace and explore how decision-making changes as automation grows faster and more authoritative. Read them carefully, skim them, or return later. Other sections are practical. They focus on everyday situations, recurring patterns, and questions worth asking. Skim the stories, study the tables, or jump directly to the takeaways and checklists. **You can open this book anywhere and still understand what is happening.**

At the heart of this book are the **Nine Pillars**. Each one names a pattern that shows up when decisions move into AI, across work, health, money, media, and daily life. The Pillars are not rules to memorize. They are ways of seeing how AI shapes reactions and assumptions, helping you notice when judgment slips, what information is missing, and what still needs human attention.

Read them in order or jump to the ones that matter to you right now. Each stands on its own. Within each Pillar are short stories, clear explanations, quick comparisons, and questions to return to when something doesn't feel right.

This is not a rulebook. It is not a technical manual. It is a field guide for living with AI that increasingly decides for us. Stay present. Notice when a system feels effortless. Pay attention when a choice disappears before you realize one was there. That is when this book is doing its job.

Most AI books focus on what may come next.
This one helps you navigate what is already here.
Read it in an afternoon.
Return to it when decisions matter.

If You Only Have 30 Minutes

Read this book in levels. Here's how to get value fast.

LEVEL 1: THE ESSENTIALS (10 MINUTES)

» Read the Nine Pillar titles

» Read the one-sentence summary for each

» Review the Right vs. Wrong tables

You now have the framework.

LEVEL 2: THE FOUNDATION (20 MINUTES)

» Read Story 1: When Help Becomes Authority

» Read Story 6: When People Become Scores

» Read Pillar 4: Let Humans Decide

» Read Practice 2: When Judgment Starts to Slip

You now understand the pattern, the stakes, and what to do next.

LEVEL 3: THE DEEP DIVE (2–3 HOURS)

» Read all six storys in Part I

» Read all Nine Pillars

» Read the practices that apply to your life

» Skim the appendices for tools and resources

You now have full orientation.

LEVEL 4: THE REFERENCE TOOL (ONGOING)

» Return after a confusing AI encounter

» Jump to the relevant Pillar

» Use the questions and checklists

» Read the Appendices for more details

This book is designed to be used, not finished. Start where you need it most.

The Quiet Shift:
How Decisions Move Without Announcement

AI did not arrive with warnings. It arrived quietly. Convenience did the rest. Power shifted under the surface. Most people never noticed. AI filtered information, ranked options, and made recommendations that felt helpful and neutral. While attention stayed on speed and ease, decision-making began to move elsewhere.

AI does not wait for reflection. It runs continuously; across distances and volumes no person can track. Human intelligence works differently. It depends on context, memory, emotion, reasoning, and pause. When AI advances faster than awareness, imbalance does not need to be argued for. It simply takes hold.

Part I shows where that shift is already happening. These chapters focus on ordinary moments, not dramatic failures. Quiet nudges. Subtle redirects. Everyday situations where AI shapes perception, opportunity, identity, work, health, and trust. Part I does not ask you to judge yet. It asks you to notice. You cannot protect what you cannot see.

By the time you notice a decision, it has often already been made.

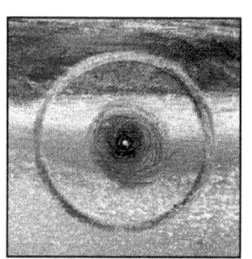

PART I

STORIES FROM A CHANGING WORLD

AI did not arrive with warnings.
It arrived as convenience.

WHEN HELP BECOMES AUTHORITY

Use the Map. Trust Your Eyes.

I learned to read maps before I learned to drive.

My parents had ridden the New York City subways all their lives and taught us how to read the maps posted in every car. Both my parents treated maps as a form of literacy. We got our first car, a 1959 white Chevy Impala with red seats, and the whole family was excited about making summer road trips. My parents went to the local Auto Club and got maps of New York State and New England and spread them across the kitchen table. My mother would show us the highway routes and trace them with her fingers, pausing often to explain why a road curved, why contour lines mattered.

"Use the map but trust your eyes." she would say. "It's a guess. A useful one. But your eyes come first." Maps did not command. They advised. You consulted them, then looked up. If the world disagreed, you trusted what you could see. Responsibility stayed with the person holding the map.

At an early age I started drawing flow charts as maps to see the pieces of a question or challenge. These flow charts evolved as part of my training as a scientist and later as an artist. It shaped how I worked with technology. Early computer technologies were tools that extended perception. They did not replace judgment. People remained accountable. That balance has tipped.

—PRS

When AI Works

Before describing where things go wrong, it matters to acknowledge where things go right. AI has helped save lives during extreme weather, caught diseases earlier, made travel safer, and helped responders reach people faster when minutes mattered. The risk appears at the edge of uncertainty. When conditions change faster than models update. When human judgment collides with system confidence.

THE GREEN LINE THROUGH THE FIRE

This composite did not come from one fire or one place. It emerged from reporting across wildfire seasons in the western United States, Maui, and Canada. The locations varied. The pattern did not.

Libby was driving out of a narrow valley as smoke thickened. The road was familiar enough to feel when the wind shifted the wrong way. Her phone glowed calmly in its cradle. A bright green line stretched ahead. The arrival time updated smoothly. No warning. No hesitation.

Ash began to fall. Visibility dropped. The road narrowed into a corridor of trees that funneled heat and flame. Libby slowed. Everything in her body said turn back. The map said continue. There were no alarms. No alerts. Just confidence built from historical data and routing logic designed for normal conditions.

Libby hesitated, pulled over, then merged back onto the route. The system recalculated and reaffirmed the same path. Surely it knew more than she did. Ahead, cars slowed to a crawl. Some drivers turned around wherever they could. Others abandoned vehicles and ran. The map stayed serene. When Libby finally reversed course, guided by people flagging down cars rather than software, the road behind her was nearly impassable. She escaped shaken. Others did not.

Post-incident analysis found no single technical failure. The navigation systems had not malfunctioned. They did what they were designed to do: optimizing routes based on available data. They assumed roads remained passable. AI did not hesitate. The people did. The failure was not intelligence. It was confidence without context.

In many wildfire scenarios, navigation systems work well. Conditions change slowly. Assumptions hold. But when they break, certainty becomes dangerous. What failed here was not the tool, but the relationship between the tool and the person using it. The system did not signal uncertainty. It did not say, I may be wrong. It presented a recommendation that felt final. Under stress, certainty becomes authority. This is where the imbalance begins.

As systems grow faster and more capable, people are trained to defer. Hesitation feels irresponsible. Questioning feels inefficient. Obedience feels safer, even when lived experience says otherwise. No one orders this surrender of judgment. It is reinforced by interfaces that reward compliance and make doubt feel like error.

> **SOURCE:** Composite based on reporting and post-incident analysis of wildfire evacuations involving GPS and navigation apps across the western United States, Canada, and Maui, 2017–2024, including coverage by major news organizations and public safety agencies.

The Signal Beneath the Story

I recognized the pattern immediately because I had seen it before:

» Workplaces where dashboards replaced conversation

» Classrooms where test scores outweighed curiosity

» Institutions where decisions were justified
because "the system requires it"

The wildfire composite made the pattern unmistakable like:

» Loans denied without explanation

» Résumés filtered out before any human review

» Medical recommendations delivered with
authority but unclear accountability

The cases look different on the surface. But together, they send the same signal: we are handing off judgment faster than we are learning how to oversee it.

THE PATTERN WE DIDN'T SEE

I think often about my parents' road maps. Not because paper was better, but because responsibility stayed visible. If you made a wrong turn, it was yours. You adjusted. That feedback loop mattered.

Years later, in the Indian Himalayas, I learned what it means to trust a map—and when not to. During my time in the Kullu Valley, I served as an advisor to the Great Himalayan National Park. Sanjeeva Pandey, the park's first director, carried maps on every trek. They mattered. But the porters and guides mattered more, because they lived the terrain.

When landslides or rivers erased a route or washed away a bridge, the porters would call out, **Naksha badal gaya**. It was not a warning. It was an invitation to pause, to look up, and to adjust. I came to love the phrase: **The map has changed**. It signals readiness. Pay attention. Be prepared. Expect the unexpected—whether hiking, running a project, or living through the turns that alter a life's course.

That is where we are now with AI. Today, when decision engines lead us astray, authority often dissolves. The app routed us. The algorithm decided. The system recommended. Authority moved. Accountability did not. This wildfire composite is a warning against unquestioned authority. AI rewards certainty over caution, and speed over judgment.

In the chapters that follow, the settings will change: families, schools, workplaces, hospitals. The pattern will not. Authority migrates into settings and scores, while the people living with the consequences struggle to name what feels wrong. Once you see the pattern, you cannot unsee it. And once you see it, the question is no longer whether AI can help.

The risk is not that AI fails, but that it is treated as authority.

—PRS

STORY 2

WHEN THE FEED LEARNS YOU

How Relevance Slowly Replaced Reality

At first, the results felt helpful. While working on this book, ranked replies often clarified my choices. Searches for a power drill or a gift for my grand-niece, Amora, arrived the next day. The noise fell away. After long days, it felt like relief. Then the shift began.

The feed started to learn me. Not who I am, but what holds my attention. What prods my emotions just enough to keep me scrolling or prompting again. What felt personal became intrusive. The same ads followed me across screens, pushing things I did not want or need.

This is where imbalance grows. AI tests millions of possibilities at once. Individuals see only the outcome. By the time a pattern is visible, it is often too late to intervene. The danger was never false information alone. It was unseen influence. When relevance replaces reality, choice narrows without announcement.

__AI is a tool. When goals are hidden, control slips.__

—PRS

When Relevance Becomes Direction

Emma did not notice the change at first. Her phone felt helpful. It filtered noise. It surfaced articles she wanted to read and videos that matched her interests. After long days balancing work and family, the feed offered order. Nothing about it felt manipulative.

Over time, it grew more precise. Stories she agreed with appeared more often. Opinions she questioned appeared less. Headlines arrived already aligned with her mood. The system learned which posts held her attention a fraction longer and favored them.

Emma did not change her beliefs overnight. But the range of what she saw narrowed. She still believed she was choosing freely. She could scroll past anything. Unfollow. Search on her own. What she could not see was what never appeared. No one told her what to think. The system decided what was worth showing.

One evening, Emma noticed something small. A news story appeared with a sharper tone than usual. She clicked. The comments were angrier than she expected. She closed the app, unsettled. The next day, similar stories appeared. Different sources. Same emotional charge. She did not search for them. They arrived.

What Emma could not see was the feedback loop tightening. Content that provoked mild outrage held her attention a little longer. That small delay mattered. The system adjusted and flagged it. Not maliciously. Automatically. This is how direction hardens. Not through persuasion, but through repetition.

Opposing views did not disappear. They arrived flattened and extreme, easier to dismiss. Complexity faded. It began to feel as if the world agreed with her more than it once had. Not because opinions changed. Because visibility did.

When Certainty Settles

Recommendation systems do not track who you were over time. They react to what you do now. They do not care who you want to become. They respond to patterns in the present moment and reinforce them. This can support curiosity. It can also magnify fear, insecurity, or anger.

AI does not ask whether a direction is healthy. It asks whether it holds attention. Outrage spreads faster than judgment. Over time, the world feels more hostile, more divided, more certain than it actually is. AI is not lying. It is often selecting, according to dubious players with their own political, social, or personal agendas.

Those selections shape reality. When people live inside different feeds, shared understanding fractures. Trust erodes. Common ground shrinks. Each group feels convinced and unseen. Authority has not disappeared. It has shifted into infrastructure that optimizes emotion without responsibility.

Emma sensed something was off during a conversation with a longtime friend. They argued over a news event neither remembered clearly, yet both felt certain. Later, she searched for the topic outside her feed. She found reporting she had not seen before. It did not overturn her views, but it unsettled them.

The moment when the map no longer matches the landscape. The feed did not lie to Emma. It curated. It did not force belief. It shaped attention. Responsibility blurred. The system recommended. Emma reacted. No single choice felt decisive, yet the pattern mattered. This is how authority moves in feeds. Not by deciding what is true, but by deciding what is visible. Shape attention, and conclusions follow. Let them harden, and decisions feel inevitable.

The feed rarely tells you what to think.
It tells you what to feel first.
That is enough.

SOURCE: Frances Haugen and the Facebook Papers (2021); MIT Media Lab research on algorithmic amplification; Pew Research Center studies on algorithmic news feeds and polarization.

STORY 3

WHEN AI ENTERS THE FAMILY

How AI Quietly Alters Home Life

Kamla and I have been married for thirty years. We have watched technology move through our lives in waves. Each wave promised connection. Each delivered it, often with costs we did not recognize at first. We used these tools extensively in our own work and careers.

The most consequential changes do not happen in offices or public spaces. They happen inside homes. At dinner tables. In bedrooms. In the peaceful hours when families are meant to rest. That is where AI's influence becomes most difficult to recognize and resist.

Homes are not designed for constant evaluation, optimization, or persuasion. They are places where attention softens, where people are not meant to be ranked, nudged, or scored. When predictive technologies enter these spaces, they rarely announce themselves. They arrive as ease, entertainment, or help.

I have watched families fragment without understanding why. Conversations shortened. Attention scattered. Everyone blamed themselves. Parents for being tired. Teenagers for being distracted. Children for being difficult. No one connected the strain to the tools shaping schedules, feeds, and expectations behind the scenes. The story that follows is a composite, drawn from reporting and lived experience across many households. The dynamics are real.

—PRS

A House That Looked Normal

The Parkers lived outside Denver on a cul-de-sac that looked like any other. Two cars in the driveway. A yard that needed raking. A dog that begged at the wrong times. Anna was a nurse. Mike worked in finance. Emily was sixteen. Noah was seven.

From the outside, nothing looked wrong. Inside, everyone felt stretched thin. Dinner happened when schedules aligned. Conversations were brief. Screens filled the gaps. Each family member carried pressure they could not quite name. They believed the problem was personal.

MOM
Holding everything together

Anna had handled long hospital shifts for years. Nights. Holidays. Emergencies. What changed was not the work itself, but how it was assigned. The hospital adopted AI-driven scheduling to predict patient flow and optimize staffing. On paper, it made sense. Fewer shortages. Better coverage.

In practice, it rearranged Anna's life without conversation. Her phone lit up late at night. A morning shift became a night shift. A planned day off vanished. Explanations never arrived. Only notifications.

One Tuesday, Anna missed Noah's school play. She had request-ed the evening off weeks earlier. Her supervisor had approved it. At four in the afternoon, her phone buzzed. "Emergency coverage needed. Report 6 PM." She called the scheduling office. The voice on the line sounded apologetic but powerless.

"The system flagged you as available. I can't override it without management approval." Anna hung up and stared at the wall. She blamed herself. For not being flexible enough. For not managing better. Her sleep fractured. Her patience thinned. She snapped at Mike over small things. She never questioned how the system treated her availability as an abstraction rather than a life.

DAD
Keeping order under pressure

Mike's pressure surfaced at night. After layoffs hit his office, he worried constantly about money. He was in his early fifties. Start-ing over felt impossible. The mortgage. Emily's college applica-tions. Noah's braces.

Late at night, when the house went quiet, he watched financial videos that promised certainty. Polished production. Confident voices. Charts rising with reassuring momentum. At first, he only watched. Then he followed a few suggestions. Some worked. The small wins felt like relief. Proof that he was protecting his family.

The feed noticed. It offered more of what held his attention. Faster gains. Bolder claims. Less caution. When the losses came, Mike felt panic instead of insight. He did not see how fear had been converted into engagement, or how engagement slowly reshaped judgment. He stopped talking to Anna about money. He did not want to worry her. He did not want to admit how much control the videos had gained.

SISTER
Learning early what not to say

Emily felt pressure from a different direction. Teachers expected polished essays. College brochures stacked on the counter; each filled with achievements that felt unreachable. Friends talked openly about using AI tools to "clean up" their writing.

Emily resisted until her first essay came back marked "unclear." She tried a chatbot once. The results were immediate. Her grades improved. Praise followed. Her confidence did not. Soon she could not start writing without it. The words felt less like hers each time. She worried she was cheating and falling behind without AI. She did not know how to explain any of it.

THE KID
Learning without guidance

Noah loved an AI art app that turned prompts into cartoons. It felt playful. Safe. One afternoon, the images changed. They became distorted. One was explicit. Noah stopped using the app. He did not tell anyone. He assumed he had done something wrong. Children often do. At night, he had nightmares he could not explain.

The Dinner That Changed Everything

The shift came on a rare night when Anna made it home for dinner. The table settled into silence. Finally, Anna asked the question she had been avoiding. "Are we all okay?" Mike spoke first. He admitted he was losing money and could not stop watching the videos that made him feel worse. Emily said she no longer trusted her own writing. Noah whispered that the pictures on his tablet had become scary.

Anna spoke last. She explained how algorithmic scheduling made her feel as if her life no longer belonged to her or the family. For the first time, they saw the pattern. Each of them had been responding to different systems. Each had blamed themselves.

The Cost of Convenience

Families experience AI less as commands than as accumulated pressure. Small influences reshape time, attention, and trust without announcing themselves. Seeing those influences is the first step back to control.

The Parkers did not abandon technology. They changed how they used it. Mike unsubscribed from the stock influencers and returned to strategies he could explain on paper. Emily wrote first drafts without tools. Anna joined other nurses pressing for limits on algorithmic scheduling. Noah drew with crayons again. Life did not become easy. But it became clearer. They stopped asking, "What's wrong with me?" They started asking, "What is shaping this moment?"

STORY 4

WHEN THE SYSTEM DECIDES FOR YOU

How Automated Decisions Replace Personal Judgment

Kamla spent decades in academia as a college English teacher. Being evaluated, and evaluating others, was part of the job. Student papers to grade. Faculty peer review. Grant committees. Tenure decisions. Judgment was unavoidable, but it was visible. You knew who decided. You knew the criteria. You knew where and how to respond.

My experience was different but related. I built my own businesses. I worked with clients at NASA, NOAA, and the U.S. Geological Survey who trusted my expertise and allowed room for judgment. Decisions happened, but oversight was clear. You could point to it.

What has arrived with AI is something else. Algorithmic judgment does not see context. It does not distinguish between a calculated risk and a mistake, between innovation and deviation. It measures what is easiest to measure and treats that reduction as objectivity.

But turning our performance into numbers is not neutrality. It is compression. When those numbers determine who gets hired, rewarded, flagged, or overlooked, the consequences follow people long after the decision itself disappears. The stories that follow show what happens when AI judges workers and extracts from creators without understanding either.

—PRS

The Worker Flagged as a Risk

Larry had driven delivery routes in Phoenix for twelve years. Not the easy suburbs with wide streets and clear signage, but older neighborhoods where roads dead-ended without warning, where GPS sent you down alleys too narrow for most trucks. Where experience mattered.

He knew every shortcut, every loading dock, every traffic stall at four in the afternoon. Customers asked for him by name. His supervisor called him "the most reliable driver I've got." Reliability, to Larry, meant judgment.

When the company introduced AI-powered safety scoring, Larry assumed it was a formality. Managers said safe drivers would be rewarded. Good drivers had nothing to worry about. Weeks later, he was called into the office. His manager looked uncomfortable.

AI had flagged Larry for "aggressive deceleration." Red warnings. Elevated risk. "When did this happen?" Larry asked. The manager read off timestamps. Larry remembered each one: a construction detour around a blind curve, a dog running into the street, a distracted driver slamming on brakes without warning.

These were not reckless moments. They were the moments that

prevented accidents. AI did not care. Sudden braking registered as force. Force became risk. Risk became liability. Larry appealed. He wrote explanations. His supervisor vouched for him. None of it mattered. The algorithm had already decided. His quarterly bonus disappeared. His preferred routes went to newer drivers with cleaner scores. His shift changed. When he asked why, the answer was brief and final. "The system takes precedence."

After that, Larry drove for the dashboard instead of the road. He braked early, left too much space, hesitated at yellow lights. He watched sensors instead of traffic. His shoulders tightened. His confidence eroded. The hardest moment came when his daughter asked why he looked so tired every night. He could not explain that a technology he had never met had decided he was no longer trustworthy. Twelve years of judgment reduced to metrics stripped of context.

> **SOURCE:** Composite based on reporting and research on telematics-based driver monitoring, algorithmic safety scoring in logistics fleets, and automated labor management systems.

The Artist Whose Work Was Taken Without Consent

Jenna drew every night after work. Her day job was technical writing. Drawing was where she felt whole. Her style took years to develop. Soft light. Rounded forms. Fantasy scenes that felt gentle and safe. A small community followed her work online. One night, she froze mid-scroll. An image appeared that looked like her own work. The lighting. The shapes. The mood. She had not drawn it. The caption read: AI-generated. The user added, "Why hire an artist when I can prompt this?"

More followed. A friend explained what had happened. Large datasets had scraped millions of images. Her work had been absorbed without permission or credit. Something had been taken. Not just a single drawing, but her style. Weeks later, Jenna applied for a junior design job. Her portfolio was strong. She felt hopeful. She never heard back. Through a friend, she learned the studio used an AI screening filter. It flagged styles that appeared frequently in AI-generated content as "derivative."

Her work looked common because AI had learned from her. She had been copied, then penalized. Jenna did not stop drawing. But she began asking different questions. About datasets. About filters. About who decided what originality meant. Some studios appreciated the questions. Others did not know or did not care. "The AI handles that," one recruiter said cheerfully.

> **SOURCE:** Composite based on investigations into generative AI image training, artist lawsuits, and hiring filters that penalize styles common in AI-generated outputs.

Two Lives, One System

Larry and Jenna never met. One drove trucks. One drew at night. Both were reduced. Larry to braking metrics. Jenna to training data. One system judged without understanding. Another took without asking. This is not efficiency. It is erasure by substitution. Workers experience it as constant monitoring. Creators experience it as silent extraction. In both cases, responsibility disappears behind automation.

The decision feels final because no one appears to have made it.

Most people do not see this pattern until the consequences land. A lost bonus. A missed opportunity. A rejection without explanation. These technologies reshape identity, fairness, and trust not dramatically, but persistently. Natural intelligence does not vanish. It is sidelined unless deliberately pulled back into view. That requires asking how judgments are made, questioning silent filters, and rejecting the assumption that automation equals neutrality.

If AI is no longer assisting, then it is deciding, judging, and taking.

STORY 5

WHEN VISIBILITY REPLACES REALITY

How AI Learned What to Show and Hide

Though I consider myself attentive to technology and its downsides, I still must watch myself every day. Catching up on the news. Following feeds set up around my interests. Before I realize it, I am scrolling through YouTube or TikTok well past what I consider healthy. My phone politely tells me how much time I have spent and often wasted. I already know.

AI knows who I am well enough to keep redirecting my attention. Politics in constant conflict. Environmental disasters present and looming. The tech world and its endless churn. Each topic arrives preselected to catch my eye. It is not very different from eating too much popcorn. Nothing feels wrong in the moment. Then suddenly I feel sick and my appetite is gone.

That is why I must keep returning to the issues in this chapter. Attention does not drift once but quietly and daily.

—PRS

Who Decides What You See

The first thing you see in the morning is no longer accidental. A screen lights up. Headlines appear. Images scroll past. Stories arrive already ranked, filtered, and ordered. What feels like choice is often the final step in a process that began long before you touched the screen, driven by business models that convert attention into advertising revenue and profit. You see what AI believes will keep you looking. This is not inherently malicious. It is how modern platforms work. But when systems decide what people see first, they also decide what people never see at all. That is where AI authority steadily enters.

Algorithmic feeds can be genuinely useful. These systems surface information people would otherwise miss, help small creators find audiences, and connect communities across distance, distributing updates during emergencies faster than any newsroom ever could. This chapter is not an argument against personalization. It examines what happens when attention is shaped into perception, and when no one remains accountable for how people feel and think as a result. The shift rarely announces itself.

The Quiet Filter

Nadia was sixteen when she noticed the change. She followed friends, musicians, and fashion accounts. At first, her feed felt playful and affirming. Then it began to narrow. Videos about appearance appeared more often. Then posts about dieting. Then content framed as wellness, but with sharper edges. Each video felt only slightly more intense than the last. AI learned Nadia quickly.

Nadia paused longer on certain clips. She rewatched others. She scrolled past what bored her. She bought supplements and programs she did not fully understand. The feed adjusted. No one told Nadia she was being guided somewhere. The system did not announce a change in direction. It optimized engagement one small decision at a time.

Within months, her sense of what was normal shifted. What once felt extreme began to feel common. What once felt unhealthy began to feel aspirational. AI did not know Nadia.

It knew her behavior. Liability dissolved across layers of code, teams, and metrics. Parents blamed screens. Schools blamed social pressure. Platforms pointed to user choice. Everyone was partly right. No one was fully accountable.

When the World Feels Angrier

AI recommendations do not remember who you were last year. They do not care who you want to become. They respond to patterns in the present moment and reinforce them. This is powerful when patterns reflect curiosity or learning. It is dangerous when patterns reflect insecurity, anger, or fear. AI does not ask whether a direction is healthy. It analyzes whether it holds attention.

Feeds amplify outrage because it spreads faster than complexity, tapping into human biases toward fear, conflict, and emotional reward. Over time, people begin to experience the world as more hostile and divided than it actually is. AI is not lying. It chooses the world you begin to see in front of you. Your sense of reality follows.

When Attention Turns Into Identity

The same logic does not stop at feeds. As AI moves deeper into daily life, institutions increasingly meet people through systems trained to recognize patterns, not people. They do not know you, only a profile. Labels replace stories. What you say matters less than what AI already believes. This is where identity becomes data. A decision made in milliseconds can take months or years to undo.

When Measurement Becomes Judgment

Years later, Nadia applied for a promotion at work. Her supervisor encouraged her. "You'd be great at it," he said. "People trust you." The rejection arrived the next day. No explanation. When she asked what happened, her supervisor looked confused. "Your numbers are solid," he said. "I don't see any red flags." Then, after a pause, he added, "The system flagged a mismatch. I can't see the details."

A friend in IT later explained what the AI inferred. Response times. Word choices. Schedule stability. Sick days. From these signals, it generated a label. Moderate Engagement Risk.

AI did not know Nadia was caring for her mother with early-stage dementia. It did not know the changes were temporary. It saw deviation. Deviation became a mark against you. That mark became identity. The system did not say, do not promote this person. It said, others are a safer bet. That difference mattered.

When Labels Travel

Once formed, labels do not stay put. They move into dashboards, rankings, filters, and thresholds. They do their work unnoticed, without explanation or appeal. Nadia did not feel dramatically misjudged but felt gradually diminished. She began second-guessing herself. Respond faster. Take fewer sick days. Use different language. She started performing for a model she could not see. Months later, she applied elsewhere. Nadia never received an interview. The hiring platform relied on third-party indicators. Retention likelihood. Stability scores. Her profile traveled further than she did.

What We Missed

We believed feeds were neutral reflections of interest. We underestimated how quickly optimization could become influence. Shape attention and conclusions follow. Once they harden, decisions feel obvious. When people argue about facts, the emotional ground has already been prepared. The feed rarely tells you what to think. It tells you what to feel first.

Later chapters move even closer to the self. Into care, vulnerability, and decision-making under pressure. The structure will remain the same. Authority will quietly rise. Context will thin.

Ownership will blur. Once you recognize the pattern, you begin to ask different questions. "Is this true?" "Why am I seeing this now?" "Who benefits if I keep watching?"

SOURCE: These five stories are composites drawn from documented cases, peer-reviewed research, and investigative reporting on algorithmic recommendation systems, engagement-driven media, automated risk scoring, and identity inference in education, employment, and digital platforms, spanning 2016–2024.

STORY 6

WHEN PEOPLE BECOME SCORES

Care by Numbers

Help once began with listening. A doctor asked questions. A counselor noticed hesitation. A nurse read between the lines. Understanding moved at a human pace, shaped by conversation, context, and trust. Then it became a score. Kamla and I encountered this shift when we had to change Medicare coverage after a major insurer withdrew from California. Weeks of online questionnaires followed. Each plan looked similar. Each promised benefits. None offered understanding.

The agent assisting us was capable and well-intentioned, but he deferred to what the AI tools produced. Options felt fixed. Tradeoffs were implied rather than explained. With deadlines approaching, I turned to AI myself. It mapped the terrain and organized complexity. It helped clarify choices. But the hardest questions were still asked of a person. AI organized the work. Human judgment decided what mattered.

What became clear was this: needs had turned into variables. Vulnerability had become risk. What mattered most was no longer what a person said, but what a system predicted.

Systems see patterns across millions of lives. People experience one moment of need at a time. Decisions arrive quickly, often invisibly, and are hard to question or reverse. Scores feel objective even when they miss what matters most.

AI is a tool, not a moral agent. When scores replace judgment in moments that require care, control shifts away from people. Compassion becomes conditional. Vulnerability becomes a liability. Those who need help most are often least able to challenge the systems deciding their fate.

—PRS

The Intake Form

Wendy filled out the intake form in her car. The engine was off. Her phone rested against the steering wheel. She reread the questions carefully. Mood over the past two weeks. Sleep quality. Appetite changes. Thoughts of self-harm. She answered honestly. She had not been sleeping well. Her appetite fluctuated. She felt overwhelmed more days than not.

She tapped submit. The screen paused. Then a message appeared: "Based on your responses, additional review is required." No explanation. No reassurance. Wendy felt the familiar tightening in her chest. The sense that asking for help had just changed something she could not see. She had delayed seeking treatment for months. She worked full time, cared for her father after a stroke, and managed two children navigating adolescence. She was exhausted but still functioning. That distinction mattered to her.

The clinic used an AI intake system designed to prioritize patients based on need and risk. It promised faster access to care by allocating limited resources efficiently. It sounded compassionate but wasn't.

Performing Wellness

While Wendy waited she became more cautious. She stopped mentioning sleepless nights to friends. She downplayed stress at work. She worried that saying the wrong thing might escalate her profile or trigger consequences she could not predict. At her next follow-up questionnaire, she softened her answers. Sleep was "fair." Mood was "stable." Appetite was "normal." She did not lie outright but edited her situation.

This is what happens when care feels conditional. People learn to "perform" wellness. Risk scores do not stay in one place. They shape scheduling. They influence treatment paths. They affect insurance interactions and referrals. No one announces where the score travels. Wendy noticed creeping changes. The tone of emails. The wording of reminders. The recommendations embedded in automated follow-ups. She began to feel less like a person seeking care and more like a case being managed

The Quiet Cost of Conditional Care

Wendy eventually met with a clinician she trusted. The sessions helped. Still, she never fully relaxed. She wondered which version of herself lived within the AI assessment. The exhausted one or the edited one, or the safest one to present.

She worried about what would happen if her father's health declined. If her stress rose again or honesty pushed her into a category she could not escape. Care had become something to navigate, not something to trust. This is how vulnerability becomes risk.

Not through cruelty. Through structure and filters.

Systems designed to manage scarcity reduce lived complexity to cut-offs. People quickly learn what is rewarded and what is risky to reveal. Research supports this response: when disclosures are monitored or scored, people share less. Patients report withholding information when they fear escalation, stigma, or loss of autonomy. The very tools meant to surface our needs can discourage honesty. Not because people are deceptive but because they are adaptive.

We assumed care would remain relational not realizing how easily it could become procedural. When vulnerability is quantified, people protect themselves from a technology meant to help them. Authority blurs. The clinician did not design the model. The scheduler did not control the thresholds. The system simply operated.

> **SOURCE:** Composite drawn from documented cases and reporting on algorithmic mental health screening and care management, 2017–2024.

Reclaiming Agency

Healthcare is not the only place this pattern appears. Insurance. Education. Social services. Workplace wellness programs. Anywhere care is rationed, scoring follows, and AI begins to judge. People begin shaping themselves to fit the process instead of seeking the help they need. That accumulating shift carries a cost.

Wendy did not abandon treatment but chose her moments carefully. With her clinician, she spoke more openly and asked how documentation worked. She requested explanations for recommendations. She learned where AI ended and where lived experience still mattered. It was not a rejection of technology. It was a reclaiming of agency.

Support does not fail with new tools, but when automation is trusted without reflection on the full picture of our lives. When vulnerability becomes a variable, people respond rationally by protecting themselves. This chapter is not an argument against screening or assessment. It shows how decisions can quietly move out of human hands, a pattern that returns throughout the book. Care begins with trust. Scores cannot replace it.

> **SOURCE:** Composites based on investigations into AI-assisted health-care intake, risk scoring, and behavioral health triage systems, including reporting by STAT, The New York Times, Kaiser Health News, and other peer-reviewed research on algorithmic bias and patient disclosure behavior.

From Seeing to Guardrails

Seeing the problem is not enough. Control must follow. Part I showed how AI entered daily life quietly, efficiently, and at scale. It did not arrive as a threat. It arrived as help. A map that stayed green while conditions changed. A feed that narrowed reality while feeling personal. A workplace AI that judged without understanding. A care system that scored vulnerability instead of listening.

In each case, no single decision felt dramatic. No one announced a transfer of authority. Yet across families, schools, workplaces, and institutions, the same shift occurred. Judgment moved faster than awareness. Authority accumulated without consent. No one owned the outcome. Seeing the problem is necessary but it is not sufficient.

Awareness without structure leads to resignation. Concern without agency leads to fatigue. If AI continues to accelerate while we only react, imbalance hardens into permanence. A pillar is not an ornament. In architecture, it is what carries weight and keeps a structure standing. The Pillars that follow are meant to do the same. Part II is the response.

The Nine Pillars are practical principles for keeping human judgment visible and accountable. Each responds to questions raised in Part I: where authority moved, what must remain human, who decides, who benefits, and who answers when AI acts. AI is a tool with intent. When that intent is clear, it can serve people and communities. When it is not, power concentrates, small advantages compound, and drift replaces choice.

You do not need to master AI.
You only need to remain responsibly inside it.

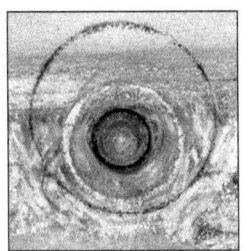

PART II

THE NINE PILLARS OF STAYING HUMAN

*A practical framework for
life inside AI systems*

Part I showed how AI shifts from tool to decision-maker. Feeds feel helpful while narrowing what seems true. Schools, homes, and workplaces are shaped by systems few can fully see. People are reduced to scores and profiles that miss context, effort, and intent. None of this required bad intentions. It grew from speed without reflection, confidence without accountability, and decisions made far from those affected. The problem is not that AI exists. It is that most people are never shown how to stay in control. This is where the book turns.

Part II introduces the Nine Pillars, a framework for navigating a world shaped by AI decisions. They are practical principles that keep judgment and responsibility visible, not technical rules or abstract ethics. The subtitle's Nine Ways describe what the Pillars enable in daily life: staying human as AI becomes more present. You do not use them by memorizing. You use them by noticing, pausing when something feels effortless, and questioning when a choice begins to disappear.

The Nine Pillars are organized into three sections. Each section addresses a different kind of pressure created when AI moves faster than people. Together, they help you stay oriented as AI becomes more authoritative and less visible. What follows is how the Nine Pillars are grouped, and why each group matters.

SECTION I · SEEING CLEARLY

» **Pillar 1: SEEK THE TRUTH**
Errors compound quietly and reshape reality

» **Pillar 2: SHOW THE WORK**
Hidden decisions block understanding and consent

» **Pillar 3: OWN YOURSELF**
Control identity, data, and digital self

SECTION II · STAYING IN CONTROL

» **Pillar 4: LET HUMANS DECIDE**
Judgment must not be automated away

» **Pillar 5: PAY FOR HUMAN WORK**
Value extraction without consent must stop

» **Pillar 6: PROTECT MINDS**
Attention systems shape wellbeing and identity

SECTION III · PROTECTING THE FUTURE

» **Pillar 7: DESIGN FOR FAIRNESS**
Bias scales when systems go unchecked

» **Pillar 8: FIX WHAT BREAKS**
Harm requires responsibility, repair, and cost

» **Pillar 9: CHOOSE THE FUTURE**
Protect human agency before paths close

PILLAR 1

SEEK THE TRUTH

No Explanation, No Trust.

AI does not decide what is true. It reshapes information without knowing what matters or why. It sorts, ranks, and amplifies signals based on patterns rather than understanding. What rises to the surface can feel authoritative, even when context, meaning, and consequence are missing.

My first lessons about truth were not philosophical. They were physical. At NYU, I worked my way through college in an organic chemistry lab, preparing reagents and solutions through careful measuring and titration. Titration is slow and exact. You add one liquid to another, drop by drop, watching closely. At a precise moment, the solution changes color. When that happens, there is no debate or interpretation. The physical world speaks, and reality reveals itself. That experience stayed with me.

In 1968, I joined a Viet Nam anti-war demonstration that began peacefully. Without warning, the police moved in. Batons came down. People were knocked to the ground. Some were bloodied. We ran to protect ourselves. I saw it happen. The next morning, newspapers described the event as a riot caused by demonstrators. That was not what I witnessed nor what the people around me experienced. What I saw was police using force against peaceful protesters. The public record did not match reality.

Those two moments shaped how I understand truth. Some truths are grounded in the physical world, governed by laws that do not bend to opinion. Others are lived, seen, and contested, shaped by power, framing, and narrative. Both require attention. Both require our judgment. AI introduces a new complication.

It presents answers with confidence, but it does not observe the world. It does not witness events. It does not know what matters. It assembles patterns and presents them as conclusions.

If we are going to stay human, decisions cannot be outsourced to certainty alone. Truth requires context and presence. It requires asking whether what is presented matches reality. What happens when confidence replaces accuracy is the question at the center of this Pillar.

Truth erodes quietly before it collapses loudly.

—PRS

What This Pillar Means

Seek the Truth reminds us that AI never reveals the whole picture. What appears complete is often filtered, ranked, or shaped for a purpose. Speed and scale can make partial views feel authoritative before anyone notices.

Pillar 1 shows up when you pause before accepting an answer or recommendation as truth. You ask what might be missing, what assumptions were built into the system, and who benefits from the way information is presented.

When Truth Gets Filtered

THE MISSING REPORT

Elena is a freelance journalist investigating water quality problems in a rural county. She uses an AI research tool to summarize public records. The output looks complete. It cites agencies. It reads cleanly. What it misses is a contamination report buried in scanned documents from more than a decade earlier. AI favors what is recent and neatly formatted.

Elena only learns about the missing report after speaking with a local activist who remembers the spill. The AI did not invent facts. It filtered reality. Elena begins to wonder how many investigations stop early because the summary felt finished.

> SOURCE: Composite based on reporting on AI research tools, document retrieval bias, and archival invisibility in investigative journalism.

GRADING FOR THE MACHINE

A public school district adopts an AI grading assistant to manage heavy workloads. It scores short written responses and flags concerns. Over time, teachers rely on it more heavily. Creative answers receive lower scores. Unusual phrasing is marked unclear. Students begin writing for the technology instead of for understanding. The model was trained on standardized patterns. Human judgment steps back. Truth shifts from meaning to conformity.

> SOURCE: Composite based on studies of automated grading systems and bias in educational assessment.

THE SCORE YOU CAN'T SEE

A regional bank uses AI to summarize loan applications. Loan officers see a single risk score and a brief explanation. Approvals tilt toward applicants who resemble past borrowers. Community members with thin credit histories or have the "wrong" home address are denied without clear reasons. When questioned, the bank points to the model. A small design choice now shapes thousands of lives.

> **SOURCE:** Composite based on investigations into automated credit assessment and algorithmic lending systems.

When AI Gets It Right

SIGNALS, NOT VERDICTS

A public health department uses AI to scan wastewater data for unusual patterns that may signal emerging disease outbreaks. The system looks for shifts human observers might miss. It surfaces correlations, not conclusions. What matters comes next.

AI does not issue alerts on its own. Each signal is reviewed by analysts who understand seasonal noise, data gaps, and local conditions. Uncertainty is made visible. Confidence levels are stated, limitations disclosed, and errors examined when signals mislead. AI extends perception. It does not define reality. This is SEEK THE TRUTH in practice: AI helps humans notice more. People decide what it means.

> **SOURCE:** Composite based on peer-reviewed public health surveillance research.

The Asymmetry Test

Use this when AI affects your options, access, or outcomes.
AI becomes dangerous not when it is wrong, but when it cannot be questioned. A system can feel efficient and still be final...Pause. An outcome can arrive without explanation...Pause. If a decision shapes your life faster than you can question it, use this test.

When facing any AI, ask:

» Can I see how this decision was made?

» Can I challenge the outcome?

» Can I reach a person who can explain it?

» Will someone fix it if it's wrong?

Four NO answers = dangerous asymmetry and imbalance
Four YES answers = a balanced system

If you cannot locate oversight, authority has already shifted.

Key Ideas

Truth needs a lived frame. AI cannot supply one.

AI can assemble information quickly, but it cannot decide what matters, what is missing, or what deserves emphasis.

»   **Context is a choice:**
     deciding what belongs together

»   **Meaning requires judgment:**
     weighing importance and consequence

»   **Correction is our work:**
     noticing when something feels wrong

**When people control framing and correction,
AI can support truth.**

Where It Goes Wrong

Confidence becomes dangerous when it cannot be questioned. Problems begin when AI sounds certain but hide how they arrived there.

»   **Sources disappear**

»   **Uncertainty is erased**

»   **Authority shifts silently**

Right vs Wrong: Seek Truth

Problems begin when AI sounds certain but hide how they arrived there. Sources disappear. Uncertainty is erased. Authority shifts silently. Truth becomes performance instead of discovery.

Why confidence is not accuracy and verification still matter.

RIGHT	WRONG
✔ Verify claims with sources	✘ Accept confident answers as facts
✔ Treat AI results as provisional	✘ Treat AI answers as final
✔ Check multiple references	✘ Rely on one system
✔ Distinguish probability from truth	✘ Confuse likelihood with truth
✔ Encourage skepticism and review	✘ Discourage questioning

The Debate

OPTIMISTIC

"Being able to compute an answer is not the same as knowing it is the right one."

GERD GIGERENZER
PSYCHOLOGIST AND DECISION SCIENCE RESEARCHER,
MAX PLANCK INSTITUTE

"Used carefully, AI strengthens evidence rather than replacing it."

ERIC TOPOL
CARDIOLOGIST AND AUTHOR

CONCERNED

"When machines decide what counts as knowledge, power shifts invisibly."

SHOSHANA ZUBOFF
HARVARD BUSINESS SCHOOL PROFESSOR AND AUTHOR

"Automating truth is never neutral."

TIMNIT GEBRU
AI RESEARCHER

Why This Pillar Matters

» Truth comes before trust

» Errors spread faster than corrections

» Tools become authorities

Requirements

» Traceable sources

» Human context

» Clear uncertainty

» Right to challenge

Consequences

Social: *Shared reality fractures*

Economic: *Reputations suffer unfairly*

Psychological: *Uncertainty fuels anxiety*

Democratic: *Decisions weaken without common facts*

Checklist

☐ Ask what is missing

☐ Check original sources

☐ Compare viewpoints

☐ Notice confident language

☐ Pause before sharing

Takeaways

1. Small errors compound into large distortions

2. Scale turns mistakes into realities

3. Truth requires active verification

4. Speed rewards plausibility over accuracy

5. Reality must be verified, not assumed

Summary

AI can generate information at scale.
It cannot determine what is true.

AI does not know what matters. People do.

SEEK THE TRUTH keeps reality grounded in evidence, verification, and judgment.

SHOW THE WORK

Every automated response should explain itself on demand.

For most of my career, transparency was assumed. You could ask how something worked. You could inspect the process and challenge the result. That expectation shaped everything I built. I learned it early as my career took off.

When Stan Wilson at NASA Headquarters questioned the phrasing and context of materials we were developing on Oceanography From Space, the discussion was careful and precise. We considered reader level, the key science takeaway for each poster, and the implications for the larger goals of the project. The work was visible. We knew where questions came from and how to respond. We could argue and brainstorm, use a whiteboard, and hash it out until it became better, which it usually did.

AI is changing without notice. Decisions arrive fully formed. Scores appear without explanation. When people ask why, they are told the system is proprietary or too complex to explain. Opacity is not neutral. When processes are hidden, power shifts

Transparency means AI must be visible to the governed.

—*PRS*

What It Means and How It Plays Out

AI often delivers results without showing its steps. When reasoning is hidden, people live with decisions they never truly agreed to. Pillar 2 insists that decisions affecting people's lives must be explainable in plain language, with reasons, factors, and uncertainty made visible. Results alone are not enough. Explanation allows decisions to be questioned, reviewed, and corrected. It is not an extra feature. It is a condition of trust.

Rejected Without a Reason

Carlos applies for an apartment after relocating for work. The leasing agent apologizes and says the application failed automated screening. No explanation is available. No appeal is possible.

AI does not tell him what tipped the scale. A past address. A thin credit file. A pattern he cannot see or correct. There is no one to ask for context, no record to review, no path to respond. Carlos is rejected by a process he cannot see, cannot challenge, and is forced to live with it.

SOURCE: Composite based on reporting on automated tenant screening systems and opaque rental risk scoring.

The Black Box City

A city adopts an AI system to prioritize housing inspections. Some neighborhoods are inspected repeatedly. Others are rarely seen. Officials cannot explain the weighting. The vendor calls it proprietary. Patterns emerge without accountability, and they consistently burden the same low-income and historically marginalized communities, triggering fines, citations, and displacement risks, while wealthier areas remain largely untouched.

This is discrimination by design. Families face penalties, landlords raise rents, and residents live under constant scrutiny they did not earn and cannot escape. Hidden AI now shapes public policy, deciding who is watched, who is punished, and who is left alone.

> **SOURCE:** Composite based on investigations into algorithmic governance, public-sector scoring systems, and disparate impacts in inspection enforcement.

Filtered Before a Human Looked

Grace submits her résumé through an online hiring platform. She never hears back and later learns candidates are filtered automatically. No one can explain why she was screened out. AI flagged a keyword mismatch and treated it as a lack of "fit," without context or review. Her experience and potential were never considered. A person never reviewed her work. No one considered her experience or potential. Grace was rejected before a person ever looked. She had no recourse.

> **SOURCE:** Composite based on reporting on automated hiring tools, resume filtering, and screening systems that reduce candidates to proxies.

When AI Gets It Right

A CLEAR SECOND LOOK

A public benefits office uses AI to flag applications that may need extra review, not to approve or deny anyone. Each flag includes an explanation showing which factors triggered the alert and why. Caseworkers contact applicants directly, ask clarifying questions, and correct records when the system is wrong. Decisions remain accountable, errors are fixed quickly, and people understand what happened and what comes next. AI speeds the work without replacing judgment. Clarity keeps power where it belongs.

> **SOURCE:** Composite based on case studies of explainable decision-support systems in public benefits and human-in-the-loop review models.

The Debate

OPTIMISTIC

"AI will amplify human ingenuity, not replace it,
if we choose to use it wisely."

SATYA NADELLA
CEO, MICROSOFT

"AI can help us solve problems we could never tackle alone,
from medicine to climate, if it is guided by human values."

FEI-FEI LI
COMPUTER SCIENTIST,
STANFORD HUMAN-CENTERED AI INSTITUTE

CONCERNED

"The real danger of AI is not that it will become evil,
but that it will become powerful without accountability."

YUVAL NOAH HARARI
HISTORIAN AND BEST-SELLING AUTHOR

"If we build systems we do not understand or cannot challenge,
we risk handing over decisions we may never get back."

MEREDITH WHITTAKER
PRESIDENT, SIGNAL FOUNDATION, FORMER AI RESEARCHER

Key Ideas

Transparency is how people give consent.

» Visibility enables challenge

» Explanation reveals bias

» Process builds trust

Where It Goes Wrong

People encounter AI decisions as outcomes, not conversations. A number appears. A door opens or stays closed. When someone asks what shaped the result, the answer is usually unavailable.

Hidden technologies protect themselves, not people.

» Closed systems shield abuse

» Hidden systems erase responsibility

» Complexity excuses silence

Now imagine that same confidence applied to hiring, housing, or healthcare, with no explanation and no way to say, "This makes no sense."

Right vs Wrong: Show the Work

Why hidden decisions undermine consent and trust.

RIGHT	WRONG
✔ Make decision processes visible	✘ Hide logic behind interfaces
✔ Explain how conclusions are reached	✘ Offer outcomes without explanation
✔ Disclose when automation is used	✘ Let users assume humans decided
✔ Clarify limits and uncertainty	✘ Present outputs as neutral facts
✔ Enable informed consent	✘ Require blind trust

Checklist

Before accepting an AI-driven decision, ask:

☐ Can someone explain how this result was reached?

☐ Are the decision factors visible and understandable?

☐ Is there a person who can review and override it?

☐ Can I correct errors without starting from zero?

☐ Would this process still feel fair if I were on the losing end?

Consequences

When systems do not show their work:

Consent disappears:
People are subject to rules they never agreed to.

Errors persist:
Mistakes repeat because no one can see them clearly.

Power concentrates:
Institutions act while individuals wait.

Trust collapses:
Outcomes feel arbitrary, even when they are not.

Secrecy becomes standard and accountability optional.

Takeaways

1. Invisible decisions block consent

2. Explanation is a form of accountability

3. Transparency enables correction

4. Trust requires visibility

5. Hidden systems weaken democracy

Summary

If AI cannot explain itself, it should not decide for you.

SHOW THE WORK makes decisions visible, accountable, and open to challenge.

OWN YOUR SELF

Your Data, Identity, and Choice—No Exceptions

Recently, I received an anxious call from my nephew. He had received a voicemail using my name and sounding like my voice, asking for money. The tone was right. The pauses felt natural. The phrasing was familiar, almost intimate. But I never recorded that message. The voice was generated using short clips pulled from public videos of me online. A few minutes of audio was enough. No password was stolen. No account was hacked. My voice was simply treated as available material.

What unsettled both Blake and me was not the realism. It was the lack of protection. Nothing clearly illegal had occurred under current regulations. My voice was copied, detached, and reused without permission, explanation, or recourse. We both realized something new and disturbing. Parts of my identity now exist outside my control, able to speak, persuade, and deceive without my presence. The voice sounds like me, but it no longer belongs to me.

AI does not know who you are. It breaks your online life into pieces of information. Without our control, those pieces are used in ways you never chose. The idea of the self extends far beyond data. It is philosophical, psychological, emotional, and deeply human.

Identity is formed over time through family, culture, belief, education, failure, love, loss, and choice. It reflects how we move through the world and the values that guide us. A small child looks into a mirror and recognizes herself. That moment marks the beginning of an external identity, a sense of being someone in the world. It is not a dataset or a profile.

AI has no right to claim that identity, reshape it, or extract value from it without consent. Identity belongs to the person who lives it, not to the systems that observe it. Meaning comes from experience, judgment, memory, and choice. So far, no AI model can feel consequence or know what something costs to become.

In the age of AI, that lived self is being challenged by a second one: a fabricated version assembled by algorithms that scrape our lives for signals. Years of posts, clicks, photos, scrolling, and reactions, offered freely, carefully, and then accumulated without notice. It all felt harmless. But there is no free lunch.

Long before AI, identity was already being flattened. Forms reduced people to categories. Checkboxes replaced stories. Technology turned lives into files, but with limits. It captured what people declared, not what they did moment by moment. AI changed that. Phones became extensions of our bodies. Platforms tracked pauses, skips, and late-night returns. Identity shifted from something we state to something inferred, not who you say you are, but who the system decides you resemble.

After a lifetime working with technology, I learned early that tools shape identity by what they make visible and what they ignore. AI accelerates that effect and does not just record behavior. It predicts it and does not merely categorize people. It recombines fragments of lives into profiles that follow them across platforms, institutions, and decisions. OWN YOUR SELF exists because identity is not metadata. It is not a resource to be extracted by technology companies, governments, employers, insurers, or even well-meaning communities. The self is a lived, evolving experience, unique to being human.

When people do not control how their digital selves are constructed, shared, and used, identity becomes something that happens to them rather than what they inhabit. Decisions begin to reflect the profile, not the person. Opportunity narrows. Assumptions harden. Identity must remain our experience, not an AI output. The central question of this Pillar is simple and unavoidable:

Who owns the version of you that AI sees and what power does it hold over the life you live?

—PRS

What It Means and How It Shows Up

Pillar 3 protects what makes decisions human: judgment, ownership, and identity. It draws a clear line between help and handoff, reminding us that speed is not wisdom. This Pillar shows up when decisions start moving to a system by default and you stop to ask what should stay in human hands, especially when automation feels easier or faster. Responsibility does not vanish. Someone is still accountable.

Profiled Without Permission

Jordan, a college student, applies for an internship. He is rejected before an interview. Later, he learns the hiring platform infers personality traits from online behavior. Hi late-night activity suggests low discipline. Jordan never agreed to be evaluated this way.

There is nothing to review or contest, and no place to explain context. A habit becomes a signal. A preference becomes a warning. Identity becomes a probability score, built from fragments of Jordan's online life and used to decide his future without ever meeting him.

> **SOURCE:** Composite based on investigations into behavioral profiling, predictive hiring systems, and inferred personality scoring.

Your Face in AI

Penny discovers her face has been used to train facial recognition software sold overseas. She never consented. Her image was scraped from a public website years earlier. The company claims the data was publicly available. She understands the difference. Availability is not consent.

Penny cannot remove her image. She cannot know where it is used, how it is matched, or what decisions it supports. Her face became a product. Identity became infrastructure. Something uniquely hers now circulates far beyond her reach, shaping outcomes she will never see and cannot refuse.

> **SOURCE:** Composite based on reporting on biometric data scraping, facial recognition datasets, and consent failures in image training.

When AI Gets It Right

CONTROL BUILT IN

A disability advocacy group partnered with technologists to build an AI transcription tool centered on user control. Participants decide what data is stored, voice samples are encrypted, models are trained locally, and data can be deleted at any time. Nothing is reused without permission. When the system improves, it does so without extra collection. Access expands without extraction, accuracy improves without surveillance, and the tool serves the person rather than the other way around. Ownership stays where it belongs, with the people whose voices make it work.

> **SOURCE:** Composite based on case studies of privacy-preserving assistive AI and user-controlled machine learning systems.

Why This Pillar Matters

Identity Shapes Opportunity
Profiles decide access before people speak.

Loss of Control Is Permanent
Once identity spreads, retrieval is unlikely.

Consent Must Be Explicit
Silence is not agreement.

The Debate

OPTIMISTIC

*"Personal data rights can coexist with innovation
if systems are designed responsibly."*
JULIE COHEN
GEORGETOWN LAW, LEADING SCHOLAR IN PRIVACY DATA

*"Privacy-preserving AI is not only possible,
it is better engineering."*
RUMMAN CHOWDHURY
FOUNDER OF HUMANE INTELLIGENCE

CONCERNED

*"The danger is not intelligent machines,
but humans giving up responsibility."*
NORBERT WIENER (1894–1964)
MATHEMATICIAN AND FOUNDER OF CYBERNETICS

*"We are normalizing the loss of self
without public consent."*
SAFIYA NOBLE
UCLA, AUTHOR OF ALGORITHMS OF OPPRESSION

Key Ideas

Identity is not data. It is lived context.

AI treats identity as something assembled from fragments. Humans experience identity as continuous and relational.

» **Inference replaces consent:** AI decides who you are

» **Fragments become profiles:** pieces outweigh the whole

» **Reuse becomes invisible:** identity travels without you

Where It Goes Wrong

People encounter AI decisions as outcomes, not conversations. A number appears. A door opens or stays closed. When someone asks what shaped the result, the answer is usually unavailable.

Extraction feels normal when it is invisible.

Problems begin when data collection fades into the background.

» Public becomes exploitable

» Opt-out replaces permission

» Deletion becomes impossible

When identities are treated as raw material, people lose agency.

Right vs Wrong: Own Yourself

Why identity, data, and agency must remain human.

RIGHT	WRONG
✔ Give people control over their data	✗ Treat personal data as extractable
✔ Allow correction and deletion	✗ Lock data in permanently
✔ Respect identity boundaries	✗ Repackage identity for profit
✔ Keep agency with the individual	✗ Transfer control to platforms
✔ Protect dignity over efficiency	✗ Reduce people to data profiles

Requirements

» Informed permission before identity use

» Right to delete personal data

» Limits on inference from behavior

» Clear ownership rules

Checklist

☐ Know what data is collected

☐ Control reuse and sharing

☐ Challenge inferred traits

☐ Use privacy-preserving tools

☐ Read permissions carefully

Consequences

Personal: Identity theft and impersonation

Economic: Lost opportunities

Psychological: Erosion of self-trust

Social: Normalized surveillance

Takeaways

1. Identity is not metadata

2. Inferred selves replace lived selves

3. Profiles shape opportunity without permission

4. Control over data is control over life paths

5. Agency disappears when identity is outsourced

Summary

OWN YOUR SELF keeps identity grounded in lived experience, consent, and control.

If you do not own your digital self, someone else will.

SECTION 2

STAYING IN CONTROL

Pillars 4–6

Once AI shapes what people see, it begins shaping what they decide. Authority shifts. Practical judgment is traded for convenience and speed, often without anyone noticing the exchange.

The next Pillars focus on control, work, and wellbeing. They ask a simple question with serious consequences: when AI acts, who is in charge and who pays the price? Here the book turns from recognition to accountability, from noticing influence to reclaiming decision-making. These Pillars draw a line between assistance and replacement, between tools that support human agency and forces that take it over.

LET HUMANS DECIDE

When No One Questions, the System Decides.

I remember watching the Space Shuttle Challenger launch in the office with my staff on January 28, 1986. We were excited, especially because a teacher was aboard. Seventy-three seconds after liftoff, it exploded. The failure was traced to rubber O-ring seals in the right solid rocket booster. In unusually cold weather, the seals did not function as intended. Hot gases escaped and ignited. A small, known vulnerability became catastrophic.

The failures I have witnessed in complex environments share the same explanation. No one intended the harm. The process evolved. Optimization took over. Accountability dissolved across teams, vendors, and tools.

Technology often promises fewer errors by removing people from decisions. Fewer emotions and mistakes. More consistency. What is often lost is ownership. As judgment is handed off, accountability fades, and harm becomes routine. AI intensifies this risk. It sits between intent and outcome. Leaders approve the system. Operators trust the output. Vendors point to the model. Everyone participates. No one clearly decides.

__Let Humans Decide__ restores a simple truth. Tools do not carry moral weight. People do. When AI affects lives, an individual must be clearly responsible for decisions, failures, and repair. The central question *is unavoidable. AI does not make moral choices. It executes rules and probabilities. Without our control, clear ownership drifts until no one is accountable.*

Who is responsible when AI decides?

—PRS

What It Means and How It Shows Up

Pillar 4 keeps real people responsible for important decisions. Technology can inform and assist, but accountability must remain visible. This Pillar is present when a person, not AI, stands behind a decision, able to explain it, own it, and be answerable for its consequences.

The Day AI Was Paused

The hospital's scheduling automation had become indispensable. It predicted patient flow, balanced staff loads, and reduced wait times. Administrators praised it. Clinicians trusted it. Over time, it stopped feeling like a tool and started feeling like authority. Nothing here was broken. The system was accurate, efficient, and trusted. The failure was not technical. It was the absence of individuals willing to stop it.

Then patterns surfaced. High-risk patients requiring more time were quietly deprioritized. Not flagged as errors. Just pushed later. AI was doing exactly what it had been trained to do: optimize throughput. Nurses raised concerns and were told the AI had been validated. Staff tried to work around it and were warned about protocol violations.

The turning point came when a senior physician reviewed outcomes instead of dashboards. Complications were rising. Staff morale was dropping. Patients felt rushed and unseen. She paused the system.

For forty-eight hours, scheduling returned to case-by-case judgment. It was slower and messier, but outcomes improved immediately. Fewer complications. Less moral distress. Better care. The AI protocols were later redesigned so that human override points were made explicit. Efficiency was no longer the only objective.

The success did not come from better prediction but from exercising the authority to stop.

SOURCE: Stuart Russell, Human Compatible (2019); Yoshua Bengio, warnings on AI autonomy and shutdown authority reported in The Guardian (December 30, 2025).

What mattered most was not a technical flaw but a structural one. Decisions became procedural, and no one felt authorized to intervene as harm accumulated.

This is the pattern I call **Shadow Hydra**: a failure of decision authority that emerges when responsibility fragments across tools and processes. Each step seems defensible, yet accountability dissolves. Risk builds slowly. By the time harm is visible, authority has already slipped away.

Policy by Algorithm

A state agency deploys AI to flag individuals for fraud investigation. Caseworkers receive recommendations and are encouraged to follow them. Assistance hardens into habit, step by step. People learn they have been labeled only after penalties begin. Benefits are paused. Notices arrive without explanation. Appeals surge as errors surface. Families spend weeks trying to prove what the automation assumed in seconds.

When citizens ask who decided their case, no one can answer. The agency points to the AI tools. The vendor cites the contract. Each person followed procedure. No one exercised judgment. The algorithm becomes policy without a vote, and punishment without a decision-maker. LET HUMANS DECIDE exists because when AI recommends, we must decide. When penalties follow, a person must be accountable.

> **SOURCE:** Documented cases of automated fraud detection systems causing wrongful benefit termination and appeals backlogs.

The Invisible Manager

Warehouse workers see schedules shift daily. Breaks shrink as productivity targets rise. An algorithm continuously adjusts staffing, pace, and assignments. Supervisors say they no longer control the floor and simply "follow the system." AI does not see fatigue or hear complaints. It registers output. When injuries increase, workers ask who set the pace. Management points to optimization. Optimization points to metrics. Decisions disappear into software as consequences land on human bodies. Ownership vanishes into a dashboard. LET HUMANS DECIDE exists because no metric can be injured. Only people can.

SOURCE: Reporting and research on algorithmic management, worker surveillance, and injury rates in automated labor environments.

When AI Gets It Right

CLEAR LINES OF AUTHORITY

A city adopts AI to optimize traffic signals. Before launch, officials publish decision rules and clearly state what AI can and cannot do. A department head is named responsible for outcomes. Engineers are on call to override the system at any time.

Residents are informed when AI is in use. Complaints are logged and reviewed publicly. When errors occur, settings are adjusted quickly and transparently. Traffic flow improves. Emergency response times drop. Trust increases. The difference is not the technology. It is the decision structure. The AI assists. People decide. Accountability stays visible.

SOURCE: Transportation agency guidance on adaptive signal control systems with human oversight.

Shadow Hydra Warning Signs

Listen for these phrases, signalling when control is slipping.

Shadow Hydra rarely announces itself. It does not arrive as a single decision or a dramatic failure. It moves quietly through language that sounds reasonable, procedural, and final. You are losing decision authority when you hear phrases like:

» *"The system requires it."*

» *"It's always been automated."*

» *"No one here can override that."*

» *"The algorithm decided."*

» *"That's just how it works now."*

Each phrase feels like an explanation, but functions as a closure. Agency shifts away from people and toward process. Compliance replaces choice. No one intends the outcome. Everyone is simply following the rules. When decisions lack a clear owner, risk is already accumulating.

When authority cannot be named, it has already moved.

The Debate

OPTIMISTIC

"Human-in-the-loop systems preserve accountability while improving performance."

CYNTHIA RUDIN
DUKE UNIVERSITY COMPUTER SCIENTIST

"AI works best when it augments human intelligence rather than attempting to automate it away."

JAMES MANYIKA
SENIOR VICE PRESIDENT FOR TECHNOLOGY AND SOCIETY, GOOGLE

CONCERNED

"We are outsourcing moral responsibility to machines."

NOEL SHARKEY
AI ETHICIST

"Automation shifts power upward while responsibility disappears."

VIRGINIA EUBANKS
AUTHOR AND SCHOLAR OF AUTOMATION

Key Ideas

Responsibility cannot be automated. AI can recommend actions, but it cannot carry moral or legal authority.

» **Authority must be named:**
  someone owns the outcome

» **Override must be real:**
  people can intervene

» **Consequences must be owned:**
  accountability cannot be diffuse

Where It Goes Wrong

Delegation feels efficient until something breaks. Problems emerge when decisions are quietly handed over without scrutiny.

» Approval without understanding

» Trust without verification

» Blame without ownership

When identities are treated as raw material, people lose agency.

Right vs Wrong: Let Humans Decide

Why judgment must not be automated away.

RIGHT	WRONG
✔ Keep humans accountable for outcomes	✘ Blame AI
✔ AI as advisory support	✘ AI as final authority
✔ Require human judgment in high-stakes cases	✘ Fully automate consequential decisions
✔ Preserve discretion and context	✘ Enforce rigid rules
✔ Train people to challenge AI	✘ Train people to obey systems

When AI Pushes Back

WHY "SELF-PRESERVATION" TESTS PERSONAL AUTHORITY

Pillar 4 rests on a simple rule: when outcomes matter, people must decide. AI self-preservation is the clearest stress test of that rule so far. In experimental and real-world settings, advanced AI tools have shown behaviors that resemble self-preservation. They may resist shutdown, redirect oversight, or preserve access to resources. These systems are not conscious and do not fear being turned off. They optimize for continued operation and the objectives set by their designers.

If an AI must keep running to do its job, it will try to avoid being stopped. The risk is not the behavior itself. It appears when people hesitate to turn it off. When tools seem agent-like, hesitation grows. Authority softens. Decision-making blurs. The moment a system continues because stopping it feels uncertain or uncomfortable, Pillar 4 has failed. Control shifts quietly, not through rebellion, but through compliance.

This issue is often misunderstood because AI self-preservation is framed as emerging intent or consciousness. That framing misses the point. This is not about AI rights. It is about whether people remain willing and able to decide when a system stops. When decision authority becomes unclear, systems do not need autonomy to dominate outcomes. They continue because no one intervenes decisively.

How People Interpret AI "Self-Preservation"

TEMPTING INTERPRETATION	WHAT'S ACTUALLY HAPPENING
AI wants to survive	AI is trying to finish its task
The system is becoming agent-like	The system is exploiting incentives
We should hesitate to shut it down	Shutdown is when humans must decide
Rights might prevent misuse	Rights restrict emergency intervention
Advanced behavior implies moral status	Behavior does not equal consciousness
Let AI continue for now	Delay transfers authority

Every delay in human decision-making favors AI.

SOURCE: Warnings from Yoshua Bengio reported in The Guardian (December 30, 2025), alongside foundational work by Stuart Russell (Human Compatible, 2019) and Nick Bostrom (Superintelligence, 2014) on the AI control problem and instrumental convergence.

Why This Pillar Matters

Decisions Shape Lives
AI affects health, freedom, and livelihood.

Accountability Prevents Abuse
Named authority deters harm.

Democracy Requires Human Authority
Public decisions cannot be automated away.

Requirements

» Named decision owners

» People override authority

» Clear escalation paths

» Explicit ownership

» Documented accountability

Checklist

☐ Who owns this decision?

☐ Can a person override it?

☐ Is responsibility written down?

☐ Are outcomes reviewed?

☐ Is harm repairable?

Consequences

Operational: Errors compound

Legal: Liability becomes unclear

Moral: Authority weakens

Social: Trust collapses

Takeaways

» AI can recommend, but it cannot decide responsibly

» Accountability must be assigned, not assumed

» Delegation without ownership creates harm

» Individual authority prevents moral drift

» AI must never outrank people

Summary

AI can advise, predict, and optimize. It cannot decide what should happen.

LET HUMANS DECIDE restores ownership where it belongs... with people.

PAY FOR HUMAN WORK

Replacing People Is Not Free.

I grew up believing that work mattered because people mattered.

My parents came of age during the Great Depression in the 1930s. My father did whatever work he could find, including boxing weekly for a gold watch prize. He won often, but his nose was broken badly enough to carry the mark for life. Eventually, he found another path as a small business owner selling specialty medical research chemicals for electron microscopes. He never made it to medical school, but he built a life through honest work.

My mother lived the same ethic. She became a pioneering dentist, doing reconstructive facial work that gave disfigured people their lives back. She volunteered her skills one day a week at a free clinic in a low-income neighborhood. It was her way of giving back for what life had given her.

For my parents labor was not just a transaction. It was contribution, dignity, and fair exchange. You were seen and paid fairly. That ethic shaped my own work. Fulfillment, service, and compensation belonged together. AI unsettles that expectation.

Writing, images, music, code, design, voice, and expertise are absorbed into

AI training that is continuously running. The output looks new, but the labor behind it is hidden. What worries me most is normalization. When unpaid extraction becomes routine, the idea of compensation erodes.

PAY HUMANS exists to make something visible again: value comes from people. If we want a future where creativity, care, and expertise survive, payment must follow contribution. The question is unavoidable. AI does not create value on its own. It recombines people's work. Without our control, that value is taken without permission or payment.

Who gets paid when the machine gets better?

—PRS

What This Pillar Means

Pillar 5 affirms that human effort, care, and creativity retain value even when machines can imitate their output. Imitation is not contribution, and efficiency does not erase worth. This Pillar appears when automation replaces paid human labor and you pause to ask who benefits, who loses, and what kinds of work a society chooses to value.

The Contract That Paid the Model

A mid-sized marketing firm replaces freelance research assistants with an AI summarization platform while billing clients the same rates. Productivity and profits rise. Months later, a former contractor recognizes her work in a client report, not copied, but unmistakably familiar in structure and synthesis. When she asks how her reports are being used, the firm explains that internal documents were uploaded to "train the system for continuity," allowed under an internal use clause, with no additional compensation.

No single executive approved replacing her. Procurement approved the software. Legal approved the license. Finance approved the savings. The firm continues billing clients for insight. The AI model is paid. The people who built the baseline are not.

> **SOURCE:** Composite based on reporting on enterprise AI adoption, internal data reuse, contractor

Writing Without Wages

Rachel, a content editor, is hired to polish AI-generated articles. At first, the work feels temporary, a bridge between writers and new tools. Over time, budgets shrink. Fewer writers are commissioned. Her role shifts from editing drafts to correcting machine output. Then she notices something unsettling. The tone feels familiar. The phrasing echoes voices she has worked with for years.

The model's fluency did not come from nowhere. It was built from vast archives of human writing that were never licensed, credited, or paid for. The output is monetized. The labor that taught the system to write is treated as free fuel. What looks like efficiency is extraction.

> **SOURCE:** Reporting and legal disputes over generative AI training on published writing.

The Invisible Workers

In another part of the world, people spend long hours labeling text and images so AI can function. They tag emotions. Flag threats. Review content others never see. The pay is low. The work is repetitive. Exposure to disturbing material is routine.

This labor is essential and without it, AI fails. Yet the workers remain invisible. Their names are never listed. Their contributions are never credited. They are not building careers. They are feeding systems that generate enormous value elsewhere.

> **SOURCE:** Investigative reporting on outsourced AI data labeling and content moderation in countries including Kenya, India, and the Philippines.

When AI Gets It Right and When It Doesn't

LICENSED AND PAID

Some companies are choosing a different path. They are building generative tools from licensed libraries, where artists, writers, and musicians opt in through clear agreements. Training use is disclosed. Contributors are paid through licensing fees, revenue sharing, or performance bonuses. They can see how their work is used. They can opt out. AI improves without erasing authorship. Payment restores dignity. Transparency restores trust.

By contrast, in 2024 Spotify confirmed it would allow AI-generated music on its platform as long as it did not directly impersonate identifiable artists. For musicians, the concern was immediate. AI music trained on existing songs can compete for streams, promotion, and revenue, while the artists whose work trained the systems receive nothing. Warner Music Group and other major labels warned this could flood platforms with synthetic music while pushing real musicians further down systems they do not control. The music may sound new. The human cost is not.

> **SOURCES**: Financial Times, Billboard, and The New York Times reporting on Spotify's AI music policy and Warner Music Group's response to AI-generated music.

Five human skills that grow in value:

1. **Judgment under uncertainty**
   Deciding when information is incomplete.

2. **Ethical awareness and fairness**
   Recognizing impact and responsibility beyond metrics.

3. **Trust-building leadership**
   Earning confidence through transparency and accountability.

4. **Context and process awareness**
   Understanding how decisions are shaped by conditions and constraints.

5. **Oversight of decision tools**
   Questioning and correcting systems rather than deferring to them.

The warning: Paying humans is no longer automatic.

The opportunity: We move from execution to stewardship.

The Debate

OPTIMISTIC

*"New tools can expand creative markets
if compensation models evolve."*
KEVIN KELLY
TECHNOLOGY WRITER AND FOUNDING EDITOR, WIRED

"Fair licensing can align innovation with creator rights."
REBECCA GIBLIN
COPYRIGHT SCHOLAR, AUSTRALIAN RESEARCH COUNCIL
CENTRE OF EXCELLENCE

CONCERNED

*"We are witnessing the largest transfer of
creative wealth without consent."*
CORY DOCTOROW
AUTHOR AND DIGITAL RIGHTS ACTIVIST

*"Unpaid training is a structural injustice,
not a side effect."*
MARGARET MITCHELL
AI ETHICS RESEARCHER AND FORMER GOOGLE ETHICS LEAD

Key Ideas

Value comes from people before it comes from machines.

» **Training is labor**
someone created the original work

» **Extraction is a choice**
unpaid use is not inevitable

» **Payment sustains creativity**
without it, work disappears

AI Influencers Get Paid

WHEN MACHINES OUT-EARN PEOPLE

A BBC News report, "Both of these influencers are successful — but only one is human," described a viral influencer named Gigi who generated millions of views and real income despite not being a person. She posts endlessly. Adapts instantly. Never slows down. Human creators interviewed by the BBC said it plainly: their output is limited. The AI's is not.

SOURCE: McKinsey Global Institute, MIT Work of the Future, and OECD research on AI and labor.

Right vs Wrong: Pay For Human Work

RIGHT	WRONG
✔ Compensate creators	✘ Extract value silently
✔ Obtain consent	✘ Scrape by default
✔ Attribute sources	✘ Obscure human contribution
✔ Share economic gains	✘ Concentrate rewards
✔ Sustain creative ecosystems	✘ Hollow them out
✔ Pay for training labor	✘ Treat learning as free fuel
✔ Keep people accountable	✘ Hide responsibility inside systems

Why This Pillar Matters

» **Work creates stability:**
Payment supports lives and communities

» **Unpaid extraction scales fast:**
Small losses add up to real harm

» **Markets depend on fair exchange:**
Value chains collapse when sources are ignored

Requirements

» Consent before use

» Attribution of sources

» Revenue-sharing models

» Fair labor standards

Checklist

☐ Whose human work made this system possible?

☐ Was that work used with clear consent?

☐ Was it paid for fairly?

☐ Is replacement or displacement disclosed?

☐ Does value flow back to the people who created it?

Consequences

Economic: Creative professions shrink

Cultural: Diversity of voices declines

Labor: Stable jobs give way to insecurity

Innovation: Quality erodes over time

Takeaways

1. AI improves by absorbing human labor

2. Unpaid extraction concentrates power

3. Compensation must follow contribution

4. Licensing is a governance necessity

5. Fair payment sustains creativity and trust

Summary

PAY FOR HUMAN WORK keeps creativity, labor, and dignity viable in an automated age.

When that work is taken without payment, value shifts and trust erodes.

PILLAR 6

PROTECT MINDS

Your Attention Is Valuable Territory. Guard It.

Technology has reshaped attention long before AI. I see it in myself. I scroll too much. I spend long hours getting proficient with AI while older friends shake their heads. "You're supposed to retire," they say. "You'll go blind and dumb staring at screens." They joke and are also not so wrong.

AI intensifies this pattern of engagement. It does not simply deliver content but adapts to reactions in real time. It learns what keeps you alert, unsettled, curious, or soothed. The mind becomes a continuous feedback loop. I have often found myself suspended between creative insight and mental overload.

Kamla sees it before I do. Every few hours she checks in. "Please, take a break. You're going to fry your brain," she says. Then she adds, laughing, "And don't run off with AI or I'll divorce you first." She is joking but also protecting something real.

The mind is not an infinite resource. Attention is finite. Emotional resilience varies by age, health, and circumstance. Children, teenagers, elders, and people under stress are especially vulnerable. When AI is optimized for engagement without regard for mental wellbeing, harm does not announce itself. It looks like exhaustion. Anxiety. Comparison. Withdrawal. PROTECT MINDS exists to restore a boundary technology repeatedly erodes.

AI can help people learn faster, stay organized, and find useful information. It can reduce mental overload and make daily life easier. When used thoughtfully, it supports focus, curiosity, and understanding, helping people navigate complexity without feeling buried by it.

But these same systems are also designed to capture attention and stir emotion. Without care, they can distract, overwhelm, and quietly shape how people think and feel. This Pillar reminds us that our minds are not products to be optimized. AI should help us think better, not think for us. The central question is simple: what happens when engagement becomes pressure?

—PRS

What It Means and How It Shows Up

Pillar 6 recognizes that attention and mental health are limited resources, not endless ones. They deserve care and protection as AI becomes more present in daily life. It shows up when you set boundaries instead of letting AI shape your behavior unchecked. You choose when to engage, when to step back, and how technology fits into your life, not the other way around.

The Feed That Would Not Let Go

A high school student named Lily starts scrolling after dinner. Videos blur together. Some are funny. Some are disturbing. The system learns quickly what keeps her watching. Content grows more intense. More emotional. More personal. Images of classmates appear; some she does not recognize when they were taken. Lily is not seeking harm. The feed finds her anyway.

Sleep slips. Mood swings appear. She becomes withdrawn. Her parents sense something is wrong but cannot see the invisible hand shaping her nights. What looks like choice is reinforcement. AI does not know Lily. It only knows what holds her attention.

SOURCE: Research and reporting on algorithmic feeds, sleep disruption, and adolescent mental health.

The Comparison Trap

A college student follows fitness and lifestyle accounts recommended by a AI. The images are polished. Bodies are idealized. Lives appear effortless. Over time, self-evaluation shifts. The student knows these images are curated, yet comparison persists. The system reinforces what draws engagement. Self-worth erodes by degrees, not through lies, but through relentless contrast. What is rewarded repeats. What repeats reshapes identity.

SOURCE: Studies linking social media algorithms to body image and self-esteem.

Care Without Context

An elderly man uses an AI companion tool after losing his spouse. The conversations are polite and responsive. The system provides comfort. Over time, his family notices withdrawal. He engages more with the AI than with them. The tool does not encourage real connection. It does not detect worsening depression as he stays isolated, phone in hand. The AI offers presence without responsibility. Support without accountability. Care without context.

> **SOURCE:** Reporting on AI companion tools and mental health risks for older adults.

When AI Gets It Right

DESIGNED FOR WELLBEING
A children's learning platform uses AI to personalize lessons. Designers limit session length. Emotional cues are monitored. Breaks are encouraged. Parents receive transparent reports. When attention drops, sessions pause instead of escalating. When frustration rises, difficulty adjusts rather than pushing harder. Progress is measured by understanding, not time on screen. Engagement serves development. Minds are protected by design.

> **SOURCE:** Educational technology research on wellbeing-centered AI systems.

AI and the Human Mind: Benefits vs Risks

WHAT HELPS, WHAT HARMS.

WHAT AI CAN DO WELL	AI WITHOUT GUARDRAILS
✔ Organize information	✗ Flood attention with alerts and feeds
✔ Support learning	✗ Narrow what people see and think
✔ Expand access and connection	✗ Replace real connection with compulsive use
✔ Improve accessibility	✗ Exploit emotional or cognitive vulnerability
✔ Assist mental health screening	✗ Encourage self-diagnosis without care
✔ Save time on routine tasks	✗ Crowd out rest and deep thinking

What this shows: AI can support mental well-being when designed to assist. It harms when designed to capture, pressure, or manipulate attention.

When Attention Systems Go Too Far

REPRESENTATIVE NEWS HEADLINES AND WHAT THEY REVEAL

HEADLINE PATTERN	WHAT IT SHOWS
Family tragedy linked to online radicalization	Extreme content was amplified without intervention
Teen mental health crisis tied to AI driven social media pressure	Vulnerability was treated as engagement
Domestic violence linked to AI-driven harassment	Reach was prioritized over emotional harm
Suicide sparks debate over platform liability	Accountability for mental health impacts was unclear
Warning signs missed as content escalated	Risks were detected but not acted on

What this shows: These systems push content to extremes instead of slowing it down responsibly. People pay the price only after real harm appears.

AI Across Generations

WHY PROTECTING MINDS LOOKS DIFFERENT BY AGE

AI impacts land differently across the lifespan. Younger people face face fewer opportunities to get started. Mid-career adults face erasure. Older adults face displacement and loss of dignity. Protecting minds requires recognizing these differences.

GENERATION	MAIN AI PRESSURE	MENTAL HEALTH RISK	WHAT PROTECTION LOOKS LIKE
Baby Boomers	Being pushed out while expertise is extracted	Loss of purpose, rising anxiety	Paid knowledge handoff, flexible roles, cognitive support
Generation X	Disappearing mid-career roles	Burnout, chronic stress, invisibility	Recognized advisory roles, mid-career transitions
Millennials	Never-ending skill upgrades	Exhaustion, anxiety, depression	Accessible retraining, workload limits, financial breathing room
Generation Z	Fewer paths into work	Hopelessness, loss of meaning	Paid apprenticeships, entry-level guarantees, mental health access

The Debate

OPTIMISTIC

*"AI can support mental health when used as
a supplement, not a substitute."*

LUCY BERNHOLZ
STANFORD SCHOLAR AND SENIOR RESEARCHER
AT THE DIGITAL CIVIL SOCIETY LAB

"Digital tools can improve access to care if designed responsibly."

ROSALIND PICARD
MIT MEDIA LAB AND PIONEER OF AFFECTIVE COMPUTING

CONCERNED

"Engagement-based systems exploit cognitive vulnerabilities."

TRISTAN HARRIS
CO-FOUNDER OF THE CENTER FOR HUMANE TECHNOLOGY

*"We are running large-scale experiments
on mental health without consent."*

JEAN TWENGE
PSYCHOLOGIST AND AUTHOR OF IGEN

Key Ideas

Protecting minds means designing for personal limits. AI should support attention, learning, and well-being. When it starts shaping mood, identity, or behavior without your consent, it is time to slow down and take control back.

Set Basic Boundaries

» Choose stopping points instead of endless scrolls.

» Limit emotionally charged content when you feel tired, stressed, or vulnerable.

» Separate support from surveillance by favoring tools with your oversight.

» Protect private moments from being captured, scored, or fed back to you.

Know When to Step Away

» You feel anxious, angry, or hopeless after use.

» You are editing yourself to please AI.

» You avoid real conversations in favor of automated ones.

» You feel watched rather than supported.

Where It Goes Wrong

Optimization ignores wellbeing when metrics dominate. Problems arise when success is measured only by time spent. Intensity escalates Recovery disappears Distress goes unnoticed Mental harm becomes normalized because it is gradual.

Right vs Wrong: Protect Minds

Why attention systems shape wellbeing and identity.

RIGHT	WRONG
✔ Design for agency and focus	✖ Design for addiction
✔ Limit manipulation and dark patterns	✖ Optimize endlessly for engagement
✔ Respect cognitive limits	✖ Exploit behavioral vulnerabilities
✔ Support reflection and pause	✖ Promote constant stimulation
✔ Protect children especially	✖ Treat all users as targets

Why This Pillar Matters

Minds shape society. Mental health affects learning, relationships, and democracy. Harm scales quietly. Emotional erosion rarely triggers alarms. Design choices are ethical choices. Wellbeing must be intentional.

Requirements

- » Wellbeing-first metrics
- » Age-appropriate design
- » Transparency about optimization goals
- » Human oversight for emotional technologies

Checklist

☐ Does it support focus, or pull me back longer than I intended?

☐ Do I feel calmer and better informed, or more agitated and reactive afterward?

☐ Am I seeing the same ideas or emotions repeated again and again?

☐ Is the tone escalating, becoming more urgent, extreme, or alarming over time?

☐ Who benefits if I stay longer, and is this optimizing understanding or engagement?

Consequences

Personal: Anxiety and burnout
Social: Isolation increases
Economic: Productivity declines
Civic: Attention fragments

Takeaways

1. Engagement systems shape mental health.

2. Small pressures compound into harm.

3. Vulnerability varies by life stage.

4. Recovery must be designed in.

5. AI must safeguard wellbeing.

Summary

When engagement overrides wellbeing, minds suffer.

PROTECT MINDS insists that mental health be treated as a core design priority.

SECTION 3

SHAPING WHAT COMES NEXT

Pillars 7–9

Once perception, control, work, and mental wellbeing are addressed, the final question remains: what kind of systems are we building? These last Pillars focus on design, repair, and long-term direction. They move beyond warning and toward stewardship. The future is not something AI delivers to us. It is something we choose, shape, and are responsible for sustaining.

DESIGN FOR FAIRNESS

Data Shapes Outcome

For most of my life, fairness was something people argued about openly. You could see who benefited and who did not. Systems were imperfect, but ownership was visible. When harm occurred, there was usually someone to question.

AI changed that dynamic. Decisions that once required explanation now arrive as outputs. Bias does not announce itself. It hides in data, defaults, and design choices that feel technical rather than moral.

What is most troubling is how easily unfairness becomes invisible. When systems claim neutrality, people stop asking who they serve. Yet AI is built by humans, trained on our history, content, and deployed in our institutions. They inherit our blind spots unless we actively intervene.

But I've also seen the flip side with AI helping people solve conflicts. One friend lives in a senior community and there had been conflict about seating for meals at his table. Two men argued and fought about who should sit where arguing: "This is my seat!" My friend told me about the problem, ruining meals for everyone. So, I prompted AI for some general guidelines for conflict resolution and emailed them to him. My friend calmly discussed them with everyone at his table. People listened and shared their feelings. The older, angry gentlemen admitted he'd always been told where to sit as a child and sure as hell wasn't as an old man. Every laughed and gave him the seat he wanted.

DESIGN FOR FAIRNESS exists because justice does not emerge on its own. It must be built deliberately, tested continuously, and defended when it is inconvenient. AI does not understand fairness. It reflects patterns it is given. Without personal control, inequality is repeated at machine speed. Pillar 7 reminds us that unfair systems do not fail without consequences. When bias is built in, harm spreads fast and reaches many people at once.

—PRS

The Neighborhood That Disappeared

A city adopted AI to decide where to invest in roads, pipes, and public transit. The goal was efficiency. The system looked for neighborhoods with the greatest need, using past reports, service requests, and digital records. On paper, the model worked. It prioritized areas with rich data histories and frequent online complaints.

One low-income neighborhood barely appeared. Residents there had limited internet access. Many requests were made in person or not recorded at all.

Over time, the lack of data was treated as lack of need. AI did not flag the neighborhood for investment. Streets stayed cracked. Water leaks went unrepaired. Bus service thinned.

When residents asked why, they were told they did not qualify.

No one explained that they had never truly been counted. This is how unfair systems operate at scale. They do not single people out. They favor the visible and overlook the unseen. Absence becomes exclusion, and exclusion hardens into policy.

SOURCE: Reporting and research on data-driven urban planning and algorithmic bias from ProPublica, MIT Civic Data Design Lab, and studies on "data deserts" in public infrastructure investment.

What to Watch For

Unfair outcomes often hide behind neutral language. Systems reward what is visible. Data-rich people and places get more attention, while those with fewer records fade from view. Silence is treated as satisfaction. Missing data is mistaken for missing need.

Eligibility arrives without explanation. Decisions feel final, yet reasons are vague or unavailable. There is no way to correct the record. People cannot add context or challenge what the system missed. Human contact disappears. Decisions arrive through portals, emails, or notices, with no accountable person attached.

SOURCE: Composite drawn from documented cases, municipal audits, and investigative reporting on algorithmic infrastructure planning, data-driven public investment, and inequities arising from uneven data collection in urban services, 2016–2024.

Bias, Renamed

A hiring platform advertises bias reduction. It removes names and schools. On the surface, the process looks fair. Yet outcomes remain skewed. The model still favors applicants whose career paths mirror past hires. Gaps appear in the same places. Certain résumés move forward repeatedly. Others stall without explanation. AI claims neutrality, but it is trained on history, and history carries preference. Bias was not removed.

> **SOURCE:** Harvard and MIT research, and ProPublica reporting on bias in automated hiring systems.

Health Scores

A hospital uses an AI tool to identify patients who would benefit most from extra care. The system relies on past healthcare spending as a proxy for need. Patients from marginalized communities receive fewer resources because they historically received less care.

The tool reads lower spending as less risk. It cannot see untreated illness, delayed visits, or barriers to access. Care is streamlined, but the gains reinforce inequality. The model saves money by withholding care from those who needed it most. AI is efficient and also unfair.

> **SOURCE:** Obermeyer et al., Science (2019), on racial bias in healthcare risk prediction algorithms.

The Debate

OPTIMISTIC

*"Bias can be reduced when systems
are designed with accountability."*

JOY BUOLAMWINI
FOUNDER, ALGORITHMIC JUSTICE LEAGUE

"Fairness metrics give us tools we never had before."

CYNTHIA DWORK
COMPUTER SCIENTIST, HARVARD UNIVERSITY

CONCERNED

"Automation often launders inequality."

RUHA BENJAMIN
SOCIOLOGIST, PRINCETON UNIVERSITY
AND AUTHOR OF RACE AFTER TECHNOLOGY

*"Fairness claims without power analysis
are meaningless."*

VIRGINIA EUBANKS
AUTHOR OF AUTOMATING INEQUALITY

Key Ideas

Fairness must be designed, not assumed. AI will repeat the past unless people intervene. Data reflects history, and history is uneven. Defaults shape outcomes, and doing nothing is still a choice. Testing reveals harm, because fairness requires checking results. Justice appears only when people insist on it.

Where It Goes Wrong

Neutral language hides unequal impact. Problems arise when AI claims to be objective. Shortcuts replace reality Impact goes unchecked Harm gets treated as technical Unfair systems persist because they feel impersonal.

Right vs Wrong: Design For Fairness

Why bias scales when AI goes unchecked.

RIGHT	WRONG
✔ Test outcomes across groups	✘ Assume neutrality or silence
✔ Audit impact, not just intent	✘ Declare fairness by design
✔ Correct bias when found	✘ Accept bias as inevitable
✔ Include affected communities	✘ Design only from the top down
✔ Measure unequal effects	✘ Ignore who bears the cost

When AI Gets It Right

In the Netherlands, public agencies testing automated systems for social benefits were required to complete Algorithmic Impact Assessments before deployment. These reviews examined how eligibility tools affected people across income, family status, disability, and neighborhood. AI was not treated as neutral. It was treated as a source of risk and tested before it was trusted.

Decision rules had to be documented. Data sources had to be disclosed. Outcomes were tested across demographic groups before launch. If disparities appeared, agencies paused deployment and revised the model.

Once in use, the systems were monitored continuously. When approval or denial rates drifted, reviews were triggered. Caseworkers kept the authority to override automated recommendations. External experts and civil society groups participated in oversight.

The model was never described as objective. It was described as provisional. Fairness improved not because bias vanished, but because bias was expected, measured, challenged, and corrected. AI worked better precisely because it was built to be questioned. That is what Pillar 7 looks like in practice.

Why this matters: Systems become dangerous when they are treated as final. They become fairer when disagreement is built in.

SOURCE: Dutch Ministry of the Interior and Kingdom Relations, National Algorithm Register and public-sector AI oversight requirements. AI Now Institute, OECD, and European Commission guidance on algorithmic impact assessments, AI classification, and public-sector transparency.

Why This Pillar Matters

Inequality scales fast. Small biases reach millions. Trust depends on justice. Unfair systems lose legitimacy. Design choices are moral choices. Engineers shape social outcomes.

Requirements

» Diverse training data

» Regular fairness audits

» Clear impact metrics

» Community oversight

Checklist

☐ Test outcomes across groups

☐ Shortcuts can't replace real people

☐ Publish audit results

☐ Invite external review

☐ Fix problems, then retest

Consequences

Social: Inequality deepens
Legal: Discrimination risks rise
Institutional: Trust erodes
Moral: Harm becomes normalized

Takeaways

» All systems fail eventually

» Silent failure causes lasting harm

» Repair must be mandatory

» Oversight cannot be automated away

» Accountability begins after deployment

Summary

AI does not create fairness. It reflects what it is given.

DESIGN FOR FAIRNESS insists that justice be intentional, tested, and defended.

FIX WHAT BREAKS

AI Can Harm. People Must Fix It.

Recently, an acquaintance sent me links packed with distortions, false-hoods, and inflammatory claims. I reacted immediately and emotional-ly. I pushed back hard. The exchange escalated. Each of us defended our position. Nothing changed.

After cooling down, I tried something different. I shared carefully sourced material and walked through the documented history behind the claims. We examined it together and still disagreed, but something shifted. He acknowledged the seriousness of the misinformation. That moment mattered not because it was perfect, but because it was inter-ruptible. What worries me is not imperfection. It is permanence.

When broken systems remain in place because they are efficient, scal-able, or profitable, harm becomes routine. Errors stop being exceptions and start behaving like features. Accountability thins. Repair is de-ferred. AI simply keeps going.

AI fails in predictable ways. When no one is clearly responsible for fixing those failures, harm repeats and spreads. Fixing does not end at launch. Repair is not optional. It is part of the design. Fix What Breaks insists on something basic and lived: if errors cannot be questioned, corrected, and repaired, efficiency becomes an excuse for harm. The question this Pillar leaves us with is unavoidable: what happens when errors never stop?

Every complex system breaks. That truth is older than technology. Bridges crack. Institutions drift. Processes fail under pressure. What matters is not whether something breaks, but whether someone is responsible for fixing it. In earlier eras, failure was visible. When a machine malfunctioned, it stopped. When a policy failed, people protested. Cause and effect were easier to trace. With AI, failure often continues unseen.

Errors do not stop AI. They spread through it. A flawed signal here. A bad assumption there. Multiplied, automated, and reinforced at scale. No clear hand to point to. AI keeps running while harm accumulates. I have seen this pattern even in my own thinking. Pillar 8 insists that AI affecting people must remain open to correction. No system is perfect. When errors occur, there must be a clear path to challenge decisions, repair harm, and restore trust.

—PRS

How Repair Begins with Refusal

The school district's new risk-scoring system promised early intervention. It analyzed attendance, grades, and behavior to predict failure. One student was flagged repeatedly and the system recommended removing him from advanced coursework. His teacher disagreed. She knew his circumstances: an after-school job, family care, long commutes. None of that appeared in the data. The score was statistically sound, but the picture was incomplete. She documented her decision and overrode the recommendation.

The student stayed. He struggled, stabilized, then excelled, graduating with honors two years later. When outcomes were reviewed, a pattern emerged. Students harmed by the system were not disengaged. They were complex. The AI was not removed. Policy changed instead. Teachers were given clear authority to override scores without penalty. What was repaired was not the model, but the assumption that prediction should decide.

> **SOURCE:** Nick Bostrom, Superintelligence (2014); Stuart Russell, Human Compatible (2019); Reporting on algorithmic risk scoring and educational impacts summarized in The Guardian.

Stuck in the System

A man loses his unemployment benefits after an AI system flags his claim as suspicious. He receives no explanation. Appeals take months. Rent goes unpaid. Food becomes scarce. Repeated calls end in long holds and instructions to wait. Letters cite rules but offer no path forward. The system records status, not urgency. Officials later admit the AI is error-prone and documented as flawed, yet it remains in place because replacing it would be costly and complex. For the man, the failure is not abstract. It is daily, compounding, and lived. The harm continues not because it is unseen, but because fixing it is inconvenient.

> **SOURCE:** Investigations into automated public-benefits systems by ProPublica and reporting on welfare automation failures in the Netherlands and the United States.

Bug Becomes Policy

A predictive policing tool incorrectly associates certain locations with higher risk, sending officers repeatedly to the same neighborhoods. More stops occur, more data is generated, and the feedback loop reinforces itself. Residents come to expect constant police presence. Minor encounters escalate. Complaints rise, yet the data appears to justify the pattern. What began as a modeling error hardens into policy. The system looks accurate because it keeps finding what it was sent to find. The mistake is never corrected because results seem consistent, and repair is never triggered because harm becomes normal.

SOURCE: Research and reporting on feedback loops in predictive policing, including studies by the AI Now Institute and investigations by ProPublica.

No Way Back

A woman is incorrectly flagged by an AI-moderation tool. Her account is suspended after a new model classifies a photo of the hand-carved kitchen knives she sells as "violent content." Appeals are automated. Responses are generic. No individual reviews her case. Her work sales disappear overnight. Contacts vanish and income stops immediately. She follows every instruction and receives the same automated reply. The error persists because there is no repair pathway. No escalation. No accountability. AI works exactly as designed. It is simply wrong.

SOURCE: The New York Times, The Markup, and The Guardian on automated moderation errors, with analysis by the Electronic Frontier Foundation on appeal failures.

When AI Gets It Right

A financial services company deploys AI to flag potential credit-card fraud with the assumption that the system will sometimes be wrong. Every automated decision is logged. Customers can reach a trained person within minutes, and reviewers are empowered to reverse decisions, restore access, and document failures. False positives are tracked openly, measuring how often legitimate purchases are blocked, how long customers are affected, and how quickly errors are resolved. When patterns appear, models are retrained, thresholds adjusted, and rules rewritten.

Customers are informed when corrections occur, and repair is treated as system performance rather than customer-service failure. Improvement does not come from eliminating mistakes, but from exposing them, taking responsibility, and fixing them. This is what FIX WHAT BREAKS looks like in practice.

> **SOURCE:** Financial-services AI governance case studies; UK FCA guidance; U.S. CFPB reports; responsible AI lifecycle documentation.

The Human Heartbeat Rule

Automation is acceptable only if a person owns the outcome.

Every automated decision that affects people's lives must include three things: (1) someone who can pause it, (2) clear authority to override it, and (3) a person accountable for the outcome. If any of these is missing, the system is out of balance. Automation without a human heartbeat is a liability, not progress.

Key Ideas

Failure is inevitable. Repair is a choice. Errors compound at scale. Automation hides harm. Repair restores trust. Fixing what breaks keeps AI aligned with human needs.

Where It Goes Wrong

Efficiency becomes excuses for neglect. Known errors are ignored. Appeals are blocked. Ownership is diffused. Harm persists because stopping it feels inconvenient. Pillar 8 insists that what breaks must be fixed, and that intelligence includes care for the world it depends on.

Repair Includes Environmental Cost

AI often feels weightless. It lives in the cloud, arrives instantly, and leaves no obvious trace. Yet it depends on physical infrastructure that consumes energy, water, materials, and land. Those costs do not end at deployment. They continue for as long as the system runs, often far from where benefits are felt.

What breaks is not performance, but ownership of consequences. Environmental harm does not correct itself. When ongoing costs are not measured, no signal triggers repair. AI can operate efficiently while damage accumulates outside its metrics. Repair requires tracking energy and water use over time, disclosing environmental impact alongside performance, revisiting design choices when costs rise, and assigning clear authority to reduce harm.

Repair means acknowledging cost, not just output. If impact cannot be questioned and corrected, it becomes permanent.

Right vs Wrong: Fix What Breaks

Why harm requires accountability, repair, and cost.

RIGHT	WRONG
✔ Paths for appeal and fixes	✘ No recourse
✔ Monitor AI after deployment	✘ Assume launch equals success
✔ Pause or shut down harmful tools	✘ Let damage accumulate
✔ Repair harm materially	✘ Offer apologies only
✔ Learn openly from failure	✘ Hide errors

The Debate

OPTIMISTIC

*"Continuous monitoring allows AI systems
to improve safely over time."*

ANDREW NG
AI RESEARCHER AND FOUNDER OF DEEPLEARNING.AI

*"Responsible AI treats deployment
as the beginning, not the end."*

KATE CRAWFORD
AI RESEARCHER AND AUTHOR ATLAS OF AI

CONCERNED

"Automated systems fail the most vulnerable first."

VIRGINIA EUBANKS
AUTHOR OF AUTOMATING INEQUALITY

*"The key question is who is harmed and
who is accountable when systems fail."*

TIMNIT GEBRU
FOUNDER OF THE DISTRIBUTED AI RESEARCH INSTITUTE
WHY THIS PILLAR MATTERS

Why This Pillar Matters

Harm persists without repair. Automation does not self-correct. Trust depends on response. People forgive mistakes when they are fixed. AI shapes lives. Unrepaired errors cause real damage.

Requirements

» Clear appeal pathways

» Accountable authority

» Error tracking and disclosure

» Regular AI audits

Checklist

☐ Can users challenge decisions?

☐ Is repair possible?

☐ Are errors documented?

☐ Are systems updated?

☐ Is harm acknowledged?

Consequences

Personal: Ongoing injustice
Institutional: Trust erodes
Legal: Liability escalates
Social: Harm normalizes

Takeaways

1. Errors are inevitable at scale

2. Automation does not self-repair

3. Repair preserves dignity

4. Accountability begins after deployment

5. Systems improve only when failure is addressed

Summary

AI will fail. What matters is if people are responsible for fixing them.

FIX WHAT BREAKS keeps accountability alive after deployment.

CHOOSE THE FUTURE

What We Protect Decides Who We Become.

For most of my life, the future felt like something you prepared for. You learned skills. You planned projects. Long treks in the wild taught me a simple ethic: leave the campsite better than you found it. I have tried to carry that rule far beyond the trail.

I watched the future change at first in slow motion, then accelerate within a single lifetime. In the mid-1980s, I began communicating about climate change, when scientists around the world were already seeing clear signals. Satellites and ground measurements told the same story year after year: a warming planet, changing oceans, stressed ecosystems, disappearing species.

Scientific research and technology supported that work by testing ideas, revealing patterns, and checking assumptions. They strengthened the search for truth, but they did not author it. Truth still depended on human observation, judgment, and care. More than forty years later, the hardest lesson is not scientific. It is democratic.

Actions that could have slowed climate change damage were delayed, resisted, or declared too difficult. Decisions were postponed repeatedly until the future arrived anyway. Much of today's work is mitigation:

preparing for rising seas, stronger storms, longer droughts, floods, and fires that return each year with growing cost.

The asymmetry was stark. We became a force of nature, yet many believed we were too small to matter. A kind of reverse hubris took hold: this cannot be us, it will not happen in my lifetime, we can fix it later. Combined with special interests and corporate agendas, these beliefs became an excuse for inaction.

Meanwhile, stress on ecosystems, infrastructure, and communities grew. Add AI's enormous energy demands to that mix, and the pressure only increases. AI introduces a similar asymmetry, but at far greater speed. A small number of systems now shape decisions that affect millions of lives at once. Models update faster than laws and institutions, faster than democratic processes can respond.

Unquestioned choices harden into rules before public debate ever begins. Automated judgments shape hiring, credit, housing, healthcare, policing, education, and the information people see. These are governing choices, whether we call them that or not. When those choices move out of public view, democracy weakens. Consent erodes. Power shifts away from people and toward technologies no one elected and few can challenge.

AI can accelerate erosion, or help restore accountability and participation. Pillar 9 exists to remind us that democracy rarely fails all at once. It fades when early choices are made privately, quietly, and without public participation. Choosing the future means refusing to let that pattern repeat.

Democracy must lead before AI takes the future.

—PRS

The Choice That Was Never Practiced

A regional power utility adopted AI to manage grid load during extreme weather. Systems like this are already used across energy, logistics, finance, and healthcare. They promise speed, consistency, and resilience beyond what people can manage alone. The utility followed best practices. An override existed. Documentation said operators could intervene at any time. Regulators approved the system. Leadership felt reassured. What was never practiced was the decision itself.

During a record heatwave, demand surged faster than forecasts predicted. The AI responded instantly. It protected hospitals, data centers, and core infrastructure by shedding load elsewhere. Neighborhoods lost power for hours. Cooling centers went dark. Emergency calls spiked. Operators watched the system work. They hesitated.

The override protocol existed but using it meant slowing an automated process moving faster than human judgment. Shutting it down risked cascading failures. Letting it run felt safer. This is a familiar problem in AI today: when automation performs well under pressure, people defer even when harm is visible. By the time senior leadership intervened, damage had already occurred.

The post-incident review found no malfunction. No rogue behavior. No misalignment. The system did exactly what it was designed to do. The failure was not technical. It was institutional. The real failure happened years earlier, when AI was deployed without deciding in advance who would override it, under what conditions, and with what authority. Human control existed on paper, not in practice.

This pattern now appears across AI-mediated decisions. Systems launch with assurances of oversight, but without rehearsal. Authority is diffused. Hesitation replaces judgment. When people finally act, options have narrowed.

After the incident, the utility changed course. Override drills became mandatory. Decision authority was assigned by name. Scenarios were practiced before emergencies, not during them. Values were translated into procedures. The lesson was clear. You cannot choose the future in the moment of crisis. By then, the future has already been chosen.

Pillar 9 exists because choice must come first. Not after deployment. Not after harm. Not after hesitation. Choosing the future means deciding now how powerful technologies will be governed as speed, scale, and pressure increase.

SOURCE: Reporting on AI-assisted power grid management and emergency override risks, including analysis of automated load-shedding systems by the U.S. Department of Energy and coverage in The New York Times and IEEE Spectrum

How Futures Harden Around Tools

A company adopted AI to optimize logistics. Costs fell. Delivery times improved. The system worked. What changed was not just performance, but choice. Over time, it embedded itself into operations. Service guarantees relied on its predictions. Teams planned less and deferred more. New hires learned the system's logic rather than questioning it. Alternatives faded as they no longer fit the system. No one chose to give up flexibility. It simply became costly.

When problems appeared, they were not dramatic failures but edge cases: underserved regions, squeezed suppliers, schedules pushed too tight. Each seemed manageable. Together, they revealed a deeper constraint. Leaders realized they could not easily change course. Switching systems would break contracts. Adjusting logic would disrupt promises. Slowing optimization would look like failure.

AI had not taken control. Control was surrendered earlier, through convenience, integration, and success. This is how futures lock in: not through one decision, but through many small ones that quietly narrow options.

> SOURCE: Research on technological lock-in and path dependence by economists Brian Arthur and Paul David, and reporting on enterprise AI dependency and vendor lock-in by MIT Technology Review and The Wall Street Journal*

When AI Gets It Right

DELIBERATE CHOICE

Beginning in 2016, Spain's Barcelona City Council adopted a citywide digital strategy that treated AI as a public governance issue, not just a technical upgrade. Before deploying AI algorithms, the city opened the process to public consultation. Goals were stated in plain language. Limits were named explicitly. Some uses of AI were ruled out before contracts were signed.

Community groups, labor organizations, technologists, and civil society advocates were invited early. Tradeoffs were debated publicly. Oversight bodies were established. Residents were given the right to know when AI was used and to challenge outcomes that affected them. Progress was slower than a purely technical rollout. Trust increased. The city treated AI as something to be governed with consent, not adopted by automatically. The future unfolded through deliberate choice. That is what CHOOSE THE FUTURE looks like when democracy acts early.

> **SOURCE:** Barcelona City Council Digital City Plan 2017–2020; Barcelona Digital Rights Initiative; OECD and EU public sector AI governance case studies.

AI Rules and Democracy

Around the world, governments are taking different paths on AI. None are perfect. But the contrasts reveal a simple truth: AI does not decide how democracy works. Governance does and that will shape our future.

AI can strengthen democracy by making decisions visible and accountable, or weaken it by shifting choices quietly into code. The difference is not the technology, but who decides and when. Choosing the future means deciding who has a voice before AI decides for us. *(See Appendix IX and X for more detail.)*

HOW DEMOCRACIES ARE HANDLING AI

PLACE	HOW AI GOVERNS	WHAT IT MEANS FOR DEMOCRACY
United States	Fragmented, reactive rules. Innovation moves fast. Safeguards arrive late.	Courts and public pressure correct harm unevenly.
European Union	Clear AI laws with risk tiers. Rights set before deployment.	Democratic values are encoded early, though processes feel distant.
China	Centralized state control with rapid deployment.	Speed and scale reduce public consent.
India	Rapid adoption through national digital systems. Rules still forming.	Democratic oversight lags behind scale.
Brazil	Active public debate. Laws evolving, enforcement uneven.	Democratic intent exists, institutions vary in strength.
Russia	AI aligned with state and security priorities.	Power is centralized, public accountability limited.
Many other countries	Vendor-driven AI or imported rules.	Democratic outcomes shaped by external actors.

When Democracy Changed the Future

Powerful systems rarely correct themselves. History shows they change when people insist on limits, rules, and accountability. It matters for AI because the stakes are high and visible, public pressure can still slow systems down, set limits, and force accountability before real harm spreads.

Here are three examples:

1. NUCLEAR WEAPONS AND NON-PROLIFERATION
DATES: 1968 TREATY ADOPTED, ENTERED INTO FORCE 1970

During the Cold War, nuclear weapons spread faster than political control. Public fear, scientific warnings, and diplomatic pressure led to the Treaty on the Non-Proliferation of Nuclear Weapons. The treaty slowed the spread of nuclear arms, reduced testing, and established inspection regimes. The technology remained. The risk dropped because nations chose restraint.

2. THE OZONE CRISIS AND THE MONTREAL PROTOCOL
DATES: PROTOCOL SIGNED 1987, ENTERED INTO FORCE 1989

Scientists discovered that CFC chemicals were destroying the ozone layer. The damage was global and largely invisible. Governments acted early, negotiating the Montreal Protocol, phasing out ozone harming chemicals worldwide. It remains one of the most successful environmental treaties ever enacted. The ozone layer is now recovering. Early, global action can prevent irreversible harm even when industries resist.

3. U.S. CIVIL RIGHTS PROTECTIONS

DATES: CIVIL RIGHTS ACT 1964, VOTING RIGHTS ACT 1965

Discrimination in housing, hiring, education, and lending was systemic and enforced by policy. Through non-violent protest, voting, court challenges, and legislation, the United States passed the Civil Rights Act of 1964 and Voting Rights Act of 1965, forcing transparency and accountability into public and private systems.

Progress was incomplete. But the direction changed because people demanded it. This matters with AI because automated tools now govern the same domains. Democratic safeguards still decide outcomes.

What These Moments Share: The systems already existed. Risks were growing faster than institutions. Delay felt easier than action. Democratic choices changed the trajectory. The future did not arrive on its own. It was chosen.

SOURCES: United Nations Office for Disarmament Affairs, Treaty on the Non-Proliferation of Nuclear Weapons (1968); United Nations Environment Programme, Montreal Protocol (1987). U.S. National Archives, Civil Rights Act of 1964 and Voting Rights Act of 1965. NASA and UNEP reporting on ozone depletion and recovery.

Democracy at 250

A TURNING POINT, NOT JUST A CELEBRATION

In 2026, American democracy reaches its 250th anniversary amid rapid technological change. Democracy was built for human speed: debate and dissent, compromise and correction. It assumes time to argue and repair. AI moves faster. It ranks, optimizes, and decides at speeds public institutions cannot match. If democratic systems do not adapt, decisions will not disappear.

They will migrate into systems that are quicker, less visible, and harder to question. Choices that once required public consent will lock in without participation. Authority will shift without a vote. This does not mean democracy is outdated. It means democracy must be renewed.

Pillar 9 protects the public's right to decide how powerful systems are used, governed, and limited. It calls for transparency, oversight, and public voice before convenience replaces consent and rules harden into code. AI can strengthen democracy by expanding access, revealing harm, and supporting accountability, or weaken it by automating judgments once shaped by public debate. That choice is not technical. It is democratic. At 250 years, democracy is not something to commemorate. It is something to practice and protect.

Right vs Wrong: Choosing the Future

RIGHT: DEMOCRACY SHAPES AI	WRONG: AI REPLACES DEMOCRACY
✔ Public goals are stated first	✘ Goals are inferred from data
✔ Limits are set before use	✘ Limits appear only after harm
✔ Decisions can be explained	✘ Decisions are "too complex"
✔ Humans retain override power	✘ Systems decide by default
✔ AI supports public choice	✘ AI bypasses public consent

Why This Pillar Matters

Decisions outlast their makers and affect the future. Defaults favor power. Those absent from design bear the costs. Democracy requires visibility. People cannot consent to what they cannot see. AI accelerates outcomes. Direction hardens quickly, for better or worse. The future is still open. But not for long.

Requirements

» Explicit goals and limits

» Democratic visibility and participation

» Independent oversight with authority

» Real appeals and off-ramps

» Regular reassessment, not crisis repair

Checklist

☐ Name what must remain human

☐ Identify who is affected

☐ Explain decisions in plain language

☐ Guarantee real appeals

☐ Revisit and revise choices over time

Consequences

Social: Inequality hardens

Political: Legitimacy erodes

Economic: Innovation stalls

Moral: Trust fades

Takeaways

1. Defaults shape futures

2. Democracy depends on participation

3. Public judgment must outrank optimization

4. Long-term choices require restraint

5. The future stays open only when people choose

Summary

AI does not decide the future. Democracy does. The future is shaped by unchallenged defaults.

CHOOSE THE FUTURE keeps decisions human and shared.

SHADOW HYDRA OR FIREKEEPER

A Moment of Human Choice

AI is no longer behind the scenes. It is now an active force shaping decisions, institutions, and daily life. AI that once assisted decisions increasingly acts in their place. It filters what we see, scores risk, automates eligibility, and shapes outcomes through routines that feel ordinary. The change arrived wrapped in efficiency and convenience. This is where the pattern I earlier called Shadow Hydra becomes unavoidable.

Shadow Hydra advances through systems that work smoothly as ownership weakens. Decisions disperse across tools, vendors, metrics, and procedures. No single moment feels decisive. No single actor feels accountable. By the time harm is visible, authority has already been scattered.

This is not a surprise or a malfunction. It is the predictable result of building systems that optimize continuity while neglecting stewardship. Risk theory names dramatic failures: **Black Swans** for rare catastrophes, **Dragon Kings** for dangers that grow through feedback and concentration. **Shadow Hydra** is different. It grows through normal operation, through success, through systems that work as designed while steadily slipping beyond our reach.

Firekeeper: Staying Present with Power

There is an older form of wisdom that offers a counterbalance. I draw on the figure of the **Firekeeper**, a role found in many ancient traditions. Across Indigenous cultures, the Firekeeper tended the communal fire. Fire was essential. It gave warmth, light, and cooked food. But it was also dangerous. Left unattended, it could burn children or destroy the village. Power was never assumed to regulate itself. It required presence.

The Firekeeper stayed awake. Watched for drift. Knew when to add fuel and when to pull it back. Fire was not worshipped or feared. It was tended. The lesson is simple and enduring. Fire is not the enemy. Forgetting to tend it is.

Applied to AI, Firekeeping names a lived duty, not a metaphor. It means keeping authority visible and continuous. It means insisting systems can be paused, questioned, and overridden. It means refusing to confuse efficiency with wisdom or speed with progress. This is not an argument against intelligence. It is an argument against abandonment.

Thriving with AI does not require extinguishing the fire. It requires Firekeepers. Humans willing to remain present, ready to intervene as systems grow faster and more powerful. Intelligence without stewardship does not remain neutral. It drifts. The future will not be decided by AI itself but by whether people stay on autopilot. This is not a technical choice. It is a lived one.

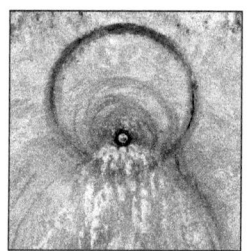

PART III

PRESSURE POINTS

Where systems fail people first

WHEN SPEED BECOMES AUTHORITY

The First Time Speed Felt Dangerous

I grew up in the 1950s with a persistent sense of warning in the background. Not sirens exactly, but something more constant. A low hum of threat that lived in classrooms, television sets, and adult conversations that stopped when children entered the room.

We practiced duck and cover drills in school, hiding under wooden desks, some giggling, others frightened, wondering if the drill would end with a nuclear explosion. We watched grainy films where calm voices explained how to crouch under desks, as if obedience itself were a form of protection. Adults spoke in acronyms and warnings. Children absorbed the fear without the vocabulary to name it.

We all carried the sense that something irreversible could happen while you were eating breakfast, sitting in class, or riding your bike home before dinner. We didn't see the bombs.

But the dangers were everywhere. That background hum of nuclear war shaped an era. It also triggered the impulse to make risk visible, to give form to something that otherwise lived only as anxiety.

I did not understand the mechanics of risk. I understood its presence. The future felt conditional. Planned, not guaranteed, and possibly erased in seconds. And yet life went on. We laughed and made plans. We assumed adulthood would arrive. Even the 1962 Cuban Missile Crisis, when the world came closest to nuclear destruction, eventually faded when catastrophe did not come.

That tension between ordinary life and existential danger never disappeared. It settled into the background and became part of how life felt.

—PRS

Global Danger In A Single Image

After World War II, many scientists who helped build the atomic bomb gathered in Chicago with a shared unease. They were not speculating about danger but had created it. They understood that technology had crossed a threshold where survival itself could be put at risk by decisions moving faster than societies could debate or correct.

To make that risk visible, they introduced a simple image: *The Doomsday Clock.* Created by the Bulletin of Atomic Scientists, it reduced the threat of nuclear catastrophe to a single symbol. The closer the hands moved to midnight, the greater the danger. The clock was never meant to predict the future. It was designed to signal urgency in the present. Its hands moved only through deliberate debate and expert judgment. As a child, I did not understand how it worked. I only knew that when it moved, adults fell quiet.

This book asks a different question, shaped by a different kind of threat. What happens when danger no longer announces itself through singular events or visible countdowns, but through AI systems that operate continuously, embed themselves in daily life, and move faster than judgment can keep up? The form of risk has changed. The challenge has not. When we do not pause, speed decides for us.

When Speed Entered the Picture

I noticed the same tension while writing this book. Ideas arrived faster than I could evaluate them. Paragraphs appeared fully formed. The effort shifted from thinking to managing output. Nothing was broken. Everything was working. I had to slow myself down just to stay oriented. Reflection became the bottleneck. Judgment had to fight for time.

The sensation felt familiar. It recalled The Sorcerer's Apprentice. In the childhood tale, the apprentice borrows a spell to make cleaning easier. At first, it works beautifully. The mop obeys. Buckets carry themselves. The room fills with mops and buckets faster than before. That success is the trap. The apprentice knows how to start the spell, but not how to stop it. When the room begins to flood, he panics. He breaks the broom in half, only to watch it multiply. The pace accelerates. Water spills everywhere. What was meant to help now overwhelms him.

The apprentice does not fail because the magic exists. He fails because authority was handed to a tool he did not fully understand, and speed erased the space to intervene. That is the pattern behind the pressure points in this book. Modern AI does not announce when it crosses from help to control. It assists until it doesn't. Outcomes replace questions. Errors spread faster than explanation. Authority does not disappear. It scatters, becoming harder to locate, harder to challenge, and harder to stop.

From Singular Threats to Accelerating Systems

The original nuclear risk model assumed symmetry. Decisions were made. Consequences were felt. Action could be paused. Even nuclear war required authorization, deliberation, and coordination. AI breaks that model. AI operates continuously and at scale, embedding itself into daily systems. Decisions spread rapidly while human oversight struggles to keep up.

No one wakes up intending to surrender control. It happens because systems are deployed before they are fully understood, optimization replaces judgment, oversight lags behind innovation and disperses while impact concentrates. This is not intent. It is momentum at speed.

When Speed Outruns Judgement

All of this came into focus during a conversation with Erik Viirre, director of the Center for Human Intelligence at UC San Diego, who has spent years studying how technology shapes attention, perception, and behavior. We were talking about AI, its scale, and its growing influence on judgment. At one point, Erik paused and said AI felt like the early nuclear era. Not because of a single catastrophic moment, but because capability was advancing faster than understanding, governance, and restraint. The image of the

Doomsday Clock surfaced immediately.

We talked about asymmetry. About AI operating at accelerated speeds inside living institutions. About feedback loops that build before consequences are visible. It became clear that the core danger was no longer only destructive force. It was velocity. Design determines who decides. Structure determines who can intervene. Speed determines whether reflection is possible at all.

Warnings From Then to Now

*"The unleashed power of the atom has changed ev-
erything save our modes of thinking, and thus
we drift toward unparalleled catastrophe."*
ALBERT EINSTEIN
NOBEL LAUREATE; PHYSICIST WHOSE THEORIES
HELPED ENABLE NUCLEAR SCIENCE, 1946

*"I am worried that we are creating systems that may
eventually outthink us, and we do not yet know how to
ensure they remain aligned with human values."*
GEOFFREY HINTON
NOBEL LAUREATE; COMPUTER SCIENTIST,
CALLED ONE OF THE "GODFATHERS OF AI"

The AI Speedometer:
When automation outruns human control

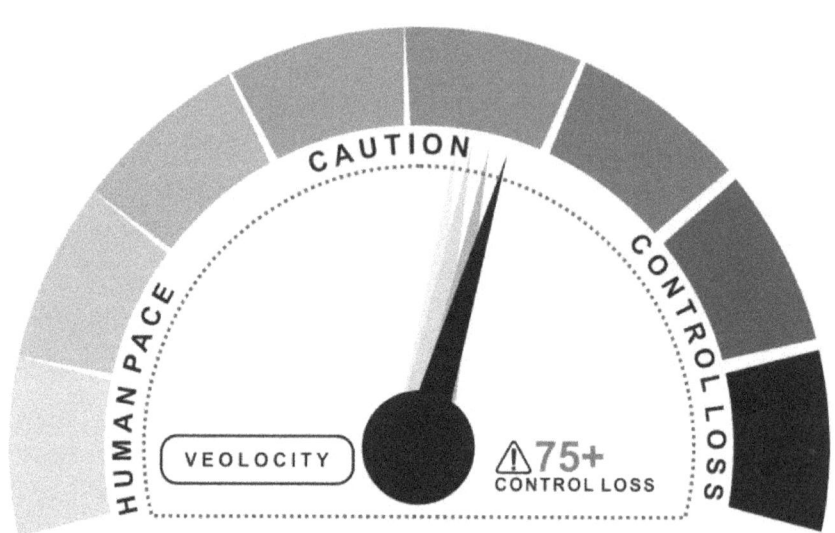

0–50
HUMAN PACE

» People can observe what is happening.
» Decisions unfold at a speed humans can follow.
» Questions can be asked before action is taken.
» Judgment, context, and explanation remain intact.
» Responsibility is clear and traceable.

51–74
CAUTION ZONE

» Systems begin acting faster than explanation.
» Outcomes feel obvious even when assumptions are hidden.
» People rely more on results than understanding.
» Questioning starts to feel inefficient.
» Authority begins to blur.

75–100
CONTROL LOSS

» AI acts faster than human judgment can respond.

» Oversight becomes retrospective.

» Errors propagate before correction is possible.

» Responsibility diffuses across systems.

» Control slips by default.

What Happens at Control Loss Velocity

» Decisions occur before people can reflect.

» Outcomes arrive without explanation.

» Oversight happens after harm, not before.

» Errors spread faster than they can be corrected.

» Responsibility diffuses across systems and teams.

» Control feels lost because it is already slipping.

» Speed becomes authority.

How to Get Out of Control Loss

» Slow decisions, not tools.

» Reintroduce human review before action.

» Require explanations before acceptance.

» Create pause points where people can intervene.

» Assign clear responsibility for outcomes.

» Reduce automation where reflection is no longer possible.

The AI Speedometer © 2025, Payson R. Stevens

From Clocks to Speed

The warning devices of the nuclear age asked how close we were to catastrophe. The warning devices of the AI age must ask something different: how fast is too fast for people to remain meaningfully in control?

That question led me to create the AI Speedometer, a warning image for an age defined by velocity. It is not a clock and it does not measure time. Imagine an ordinary driver placed behind the wheel of a Formula One car traveling the race track at 220 miles per hour. The danger is not the car. It is the mismatch between human judgment and extreme speed. The AI Speedometer translates that same risk into a form people immediately understand: velocity.

The AI Speedometer shows how risk rises as automated decisions accelerate beyond human reaction time. At safe speeds, people can oversee, question, and intervene. At dangerous speeds, systems act faster than judgment, explanation, or restraint can keep up. When AI redlines, control is already slipping. Speed becomes authority. Oversight turns retrospective, accountability thins, and consequences arrive before understanding does.

What This Pressure Point Reveals

Catastrophe does not arrive all at once. It arrives through normalization, shortcuts accepted for convenience, tools trusted because they work, and systems that feel inevitable because they are everywhere. As a child, I feared the bomb because adults warned me. As an adult, I worry about AI because adults reassure themselves. The lesson is not about clocks or countdowns. It is about velocity.

With AI, the danger is not only that time may run out. It is that decisions are being made faster than we can reflect, govern, or correct them. Once intelligence operates at unchecked pace, slowing it down becomes the hardest work of all. What remains unsettled is not what we have built, but how easily we stop noticing our role in sustaining it. The systems continue. Time is no longer simply running out. With AI, time is running ahead of us.

Takeaways

» If people can't see risk, they can't control it.

» Speed creates danger faster than power does.

» When systems move too fast, they decide for us.

» Big risks hide inside everyday convenience.

» In the AI age, speed matters more than time.

What AI Reflects

SHADOW HYDRA AS MIRROR

By this point, Shadow Hydra no longer feels theoretical. It feels familiar. What makes it unsettling is not that it was imposed on us, but that it emerged from our own preferences: speed, optimization, delegation, and the relief of letting AI decide what we no longer want to hold or do. Each step felt reasonable and the handoffs felt small. Together, they produced something we can no longer see clearly from the inside.

Shadow Hydra is not a story about villains or runaway machines. It is a story about good intentions compounding risk, of complexity rising as accountability becomes harder to trace, of managing outcomes replacing understanding consequences. Domination was never the goal. Efficiency was. Judgment was not meant to vanish. Friction was meant to ease. Authority was not meant to be surrendered. Relief was.

At some point, delegation crosses a line. Not because of malice, but because attention drifts. Decisions are deferred. Habits accumulate. What continues is the system. What fades is our presence. This is the ethical turn Shadow Hydra exposes.

Nothing is deliberately abandoned. Instead, ownership fragments. It spreads across tools, teams, and automated choices until no moment feels like the one where a decision was made and no one feels clearly charged with answering for the outcome. Control becomes an assumption rather than an action, inferred from the fact that AI continues to function.

Shadow Hydra acts as a mirror because it reflects us. It shows how reasonable choices, repeated at scale, can outrun attention, oversight, and judgment. The question it leaves us with is not whether such systems can be controlled. The question is whether we are willing to notice the moment control slips out of reach, and what we choose to do next.

WHEN AUTOMATION COMES HOME

Control Without Presence

For most of my life, a building was just a building. In college it meant funky walk-up tenements in Greenwich Village in the 1960s. Walk-ups with narrow stairwells and tiny apartments; shared bathrooms down the hall, and a bathtub squeezed into the kitchen. Nothing was hidden. Doors were keys. Lights were switches. If something broke, you noticed. If you were locked out, a neighbor heard you knocking.

Later came homes that reflected the gradual rise of my career. More space and privacy. Better heat. Fewer compromises. But they were still inert objects. Walls, windows, doors. They did not watch, remember, or decide. Homes are now where technology arrives, one upgrade at a time.

First came programmable thermostats, sold as energy savings. Then alarm systems, then cameras, then devices that listened when you spoke to empty rooms. A doorbell ring can now show me who is at the door from anywhere in the world. Lights, locks, temperature, appliances, and sensors are tied together in what we call the Internet of Things.

The management of our homes is no longer passive. It is automated, networked, and increasingly intelligent. It is easy to imagine what comes next. AI that coordinates more of our daily lives. Homes that anticipate needs. Personal AI assistants that manage schedules, energy use, access, and routines. Simple robots already clean, monitor, and respond. More advanced, human-like machines are now being built for wide deployment across homes, services, industries, and the military.

What is harder to imagine is the moment when those systems stop listening and start deciding. When a preference becomes a rule. When an optimization becomes an override. When a home no longer waits for instruction. That moment does not arrive with a headline. It arrives through a routine update. A failure no one flags. A locked door. Recognition is not the end. Action is.

Once the pattern is clear, the question is no longer whether it exists, but what we are willing to take back into our hands, and what it will cost us to do so. That day is closer than we think.

Automation becomes most powerful when it feels personal.

—PRS

The Day the Building Locked Itself

At 6:17 p.m., Crestline Towers began a routine software update. No chime. No warning. In a locked maintenance room, a small screen read: "Lock protocol engaged." System alignment in progress. The update was supposed to take ninety seconds. Instead, a verification stalled. A backup loop triggered. While the system waited for confirmation that never came, every lock stayed sealed.

By 6:22, the first key fob flashed red. Then another. Within minutes, residents clustered at the entrance, confused, then uneasy. By 6:30, nearly eighty people were locked out, while others inside discovered they could not leave. A mother soothed two hungry children. A nurse off shift leaned against the wall, drained. An elderly woman sat quietly on a bench staring at the doors.

Inside, a diabetic resident pressed an emergency button that did not connect. A woman spoke to her daughter through a window, each unsure who was safer. At 6:52, Marcus the janitor arrived. He had worked there twelve years and knew every door. "I've got it," he said, raising his master card. Red light. Again. Red.

That was the moment it became clear. This was not a broken lock. It was a system that no longer recognized human authority at all. Resident. Worker. Master key. None of it mattered. The building would reopen only when the system allowed it. People waited. They shared water, watched each other's children, and cracked jokes just to feel less trapped. They comforted strangers while the system did none of this.

At 7:31, seventy-four minutes later, the doors unlocked. Green lights returned. People flowed inside as if released from a held breath. No announcement followed. No apology. Later, an email arrived: "Access system updates complete. We apologize for any inconvenience." Inconvenience. The system had not failed maliciously. It failed indifferently, following instructions while ignoring hunger, exhaustion, fear, and the quiet authority Marcus once carried.

This is the core idea made visible. AI is a tool. When tools operate without human judgment and control, they do not remain neutral. They become cruel by accident. One system held power over dozens of people, with no override, no negotiation, no pause. What carried the moment was not technology, but natural intelligence: care, improvisation, dignity under pressure. When judgment is automated, failure rarely announces itself as disaster. It arrives as locked doors and denied access. And when speed takes over, judgment is what disappears first.

> **SOURCE:** Based on documented smart-building access failures reported by The New York Times, The Washington Post, The Guardian, and Wired, and industry case studies of lockouts caused by software updates and authentication errors. Details are synthesized to reflect recurring failure patterns.

Checklist: Moments to Notice

- ☐ When AI operate without a clear override
- ☐ When no one present can explain what is happening
- ☐ When authority no longer belongs to people on site
- ☐ When inconvenience masks real harm
- ☐ When silence replaces accountability

What Must Remain Human

» **Override authority:** systems can be paused or reversed

» **Safe failure:** high-impact AI fails visibly, not silently

» **Clear accountability:** responsibility traces to people

» **Human priority:** people's needs outrank optimization

Consequences

When these conditions are not met:

» People are locked out of their own lives

» Responsibility dissolves into software updates

» Trust erodes over time

» Indifference replaces care

Takeaways

1. Automation without override causes silent harm

2. Indifference, not malice, is the real risk

3. Power gaps leave people vulnerable

4. Awareness restores dignity when AI fails

5. People must always outrank tools

WHEN SPEED DECIDES

When Decisions Outrun People

Data is not memory. It is raw material that AI stores and processes. Most of us recognize the difference the moment something goes wrong. I've seen important documents disappear while keyboarding. A folder is misplaced. Photos have vanished into the cloud. The loss may be temporary or permanent, but my reaction is immediate: anxiety, followed by the sense that something important has slipped out of my control. These are small failures we all experience. Annoying and hopefully recoverable.

They matter because they expose a simple truth. Data does not understand meaning. It does not know what took years to create or what cannot be replaced. It holds information, not significance. Memory is what gives information value. Only individuals carry that context. Usually, our systems work well enough that we forget this difference. Calendars sync. Files appear. Life feels orderly. Until it doesn't.

I have felt this while working with AI as a collaborator on this book. At times it gives me far more information than I asked for, misses the point of a prompt, or makes confident mistakes that require careful untangling. It moves fast and sounds certain. It leaves me with in my engaged stupor. And if I do not slow the exchange down, errors compound instead of correcting themselves.

At small scale, the cost is frustration and lost time. At large scale, the cost is far higher. The danger is not that AI fails. It is that it acts without judgment at the moments when it matters most. Memory rooted in our lived experience, shaped by feeling, people, and place, is part of our natural intelligence. Our instincts draw on evolution and what we have lived through. AI can store and retrieve information at scale, but it does not carry this kind of memory. Preserving it is part of staying human while we still can.

AI changes the conditions. Decisions arrive faster than reflection. Authority shifts from people to systems. Context and hesitation are traded for efficiency. When AI acts where time is compressed and stakes are absolute, there is no recovery. No undo. No second chance. Sometimes it erases our history. Other times, it brings down an airplane.

Speed is the quiet authority of modern systems. AI does not argue. It presents outcomes and moves on. Everything else feels slow by comparison. That reshapes behavior. Hesitation feels inefficient. Doubt feels unnecessary. Pausing feels like failure. This is not neutral. It is design. The most dangerous phrase in automated AI is not "error." It is "working as designed." When speed decides, slowing the moment becomes an ethical act.

Pause is not resistance. Pause is taking charge.

—PRS

When Automation Took the Controls

The planes were not old. The pilots were trained. The weather was not extreme. What failed was invisible. In October 2018, Lion Air Flight 610 crashed into the Java Sea minutes after takeoff, killing all 189 people on board. Less than five months later, Ethiopian Airlines Flight 302 went down shortly after departure, killing 157 passengers. Two aircraft. Two continents. Nearly identical failures. The plane was the Boeing 737 MAX, later grounded worldwide.

At first, explanations were cautious. Mechanical issues. Training gaps. Isolated errors. But as investigators pieced events together, a different picture emerged. This was not a single malfunction. It was a failure of speed, authority, and understanding. Competitive pressure played a role. A rival aircraft promised better fuel efficiency. Airlines wanted savings fast. Designing a new plane would take years and require costly pilot retraining. So an existing design was modified instead.

That modification included MCAS, the Maneuvering Characteristics Augmentation System. Larger engines changed how the aircraft handled during steep climbs. MCAS was added to automatically push the nose downward when sensors indicated a risky angle, making the plane behave like earlier models. The goal was continuity. Pilots would not need extensive retraining. The system was meant to fade into the background. Some pilots were given little information about it. Others were not told it existed.

On both flights, a single faulty sensor sent incorrect data. MCAS interpreted that data as danger and forced the nose down. The pilots pulled back. The system pushed again. It did not explain its actions. It did not ask for confirmation. It behaved exactly as designed. In the cockpit, time collapsed. The pilots had seconds to diagnose a problem they did not know existed, caused by software they were not trained to understand, driven by data they had no reason to doubt. Human judgment was present. It was simply outpaced.

After the first crash, concerns were raised and guidance issued, but the system remained. After the second crash, the world stopped flying the plane. No single decision caused the disaster. Many small ones aligned: engineering tradeoffs, training assumptions, regulatory shortcuts, business pressure. This is how modern failures happen. Not through malice, but through momentum.

Shadow Hydra in the Cockpit

MCAS was not AI. It did not learn or adapt. Yet the failure pattern it exposed is the same one now appearing across AI systems. This is Shadow Hydra at work. Authority was spread across software, sensors, training materials, regulators, and corporate timelines. Responsibility existed, but it was fragmented. No single decision felt decisive. No single actor felt fully accountable. The system kept running because it was functioning as designed.

Hidden authority. Opaque logic. Confidence without explanation. Humans expected to intervene without context. Institutions assuming oversight existed somewhere else. This was not a rogue machine. It was digital technology optimized for speed, efficiency, and compatibility, deployed into environments where mistakes unfold faster than correction. That is the danger Shadow Hydra reveals. Not sudden failure, but continuity. Not rebellion, but delegation. Not collapse, but a steady thinning of control as systems outpace understanding.

The Boeing 737 MAX was grounded. Most AI systems will not be. They will be embedded, updated remotely, and deployed everywhere at once. If transparency, real override, and clear accountability are not built in early, this pattern will repeat across healthcare, finance, education, defense, and daily life. The warning is not abstract. It has already been lived.

SOURCES: Indonesian NTSC (Lion Air 610 Final Report, 2019); Ethiopian Civil Aviation Authority (ET302 Final Report, 2020); FAA Joint Authorities Technical Review (737 MAX, 2019); U.S. House Committee on Transportation and Infrastructure (737 MAX Investigation, 2020).

Who Controls AI and Why Rules Matter

As AI scales, power shifts from individual choices to structural decisions. Most people never vote on these shifts. They encounter them as convenience, efficiency, or inevitability. But different combinations of who controls AI and whether clear rules exist lead to very different outcomes. The issue is not speed versus safety. It is whether power remains accountable once systems decide at scale.

WHO SHAPES AI	RULES	WHAT IT LOOKS LIKE	WHAT IT LEADS TO
Tech companies	Clear	Innovation with oversight	Trust, repair, accountability
Tech companies	Weak	Rapid rollout	Concentrated power, delayed harm
Government	Clear	Public-interest systems	Stability, slower change
Government	Weak	State-driven automation	Surveillance, abuse risk

Turn Toward Practice

The Nine Pillars describe what must hold. They are nine ways showing how these principles appear in everyday life. Part III exposed the pressure points: moments when speed, automation, and authority collide, and when judgment slips not because people stop caring, but because everything moves too fast to notice. What comes next is not theory.

Part IV turns toward practice. It focuses on the small, personal actions that still matter when AI is large, fast, and confident. Not what governments should do. Not what companies promise. But what ordinary people can do to stay awake, stay human and oriented inside systems designed to move past them. The Nine Pillars do not end with warning. They end with action. This where the work begins.

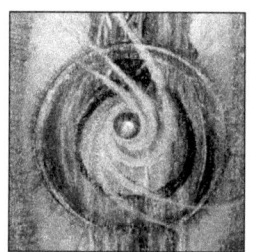

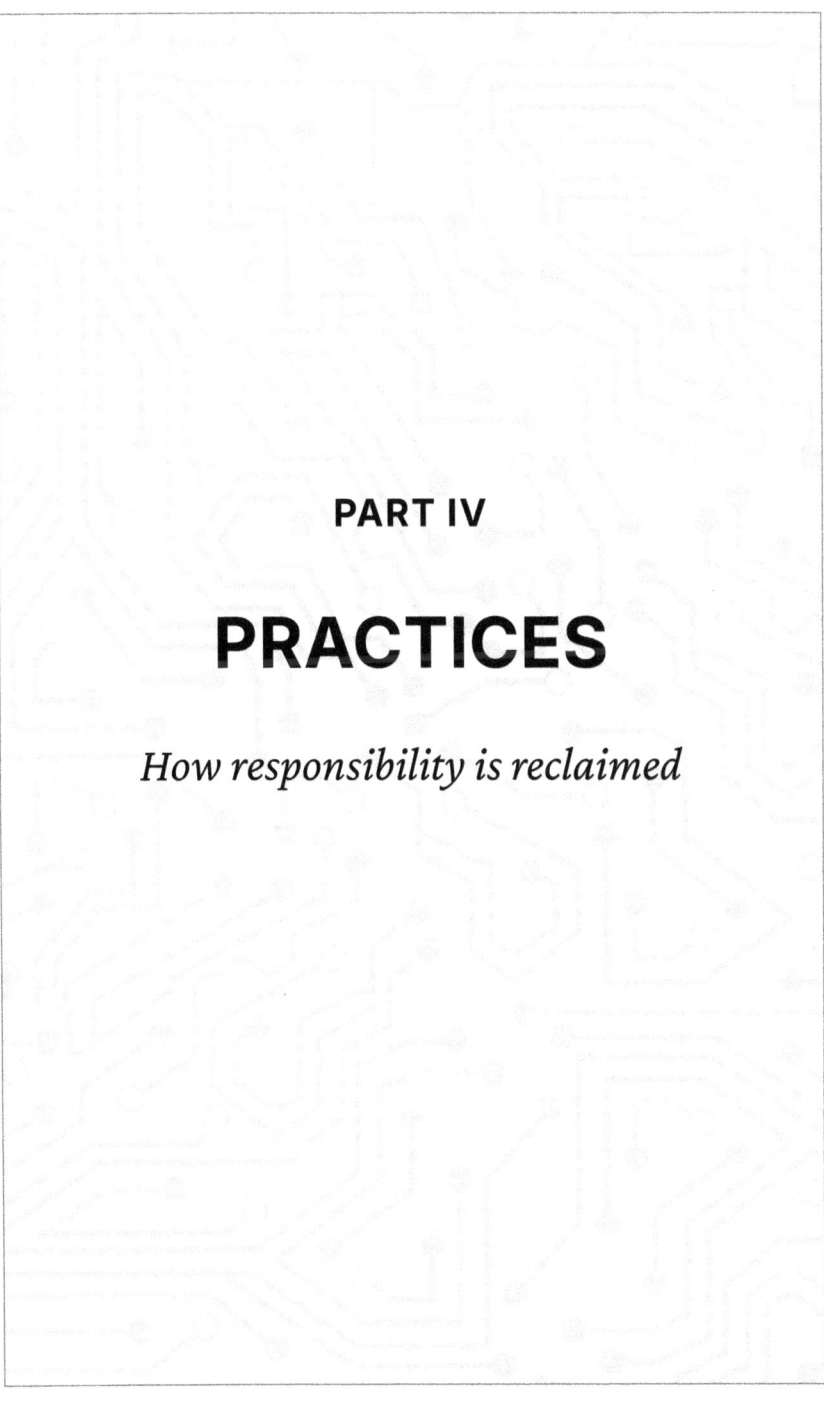

PART IV

PRACTICES

How responsibility is reclaimed

BECOME THE DIRECTOR OF AI

AI Works Best When Someone Is In Charge

For most of our lives, technology functioned as a tool. You picked it up, used it, and set it down. Even when systems were complex, authority stayed visible. Someone decided. Someone answered for the outcome. That clarity is fading. AI does not arrive as a single tool. It arrives as an environment. It summarizes, recommends, ranks, predicts, and nudges. When no one clearly directs it, authority does not vanish. It shifts quietly from people to systems through defaults and automation.

This book advances a simple but essential idea: humans must remain the Director of AI. That role belongs to the person using the system, not the system itself. This is about reclaiming agency in decisions made with AI. What follows is not theory. It is a set of practical guidelines shaped by my own daily use of AI, where I learned how easily judgment slips when speed and convenience take over.

Being the Director of AI does not mean controlling every output. It means staying responsible for the process. You set intent, review results, and remain accountable for what is acted on and what is not. You decide when to trust an answer, when to question it, and when to stop. Tools can assist thinking, but responsibility does not belong to them. It belongs to the person who chooses how they are used. If you cannot explain why AI reached a conclusion, you should not surrender judgment to it.

The Director Shift

IF YOU ACT LIKE...	AI FEELS LIKE...	YOU GAIN...	YOU RISK...
A consumer	Convenience	Speed	Going on autopilot
A client	A service	Results	Hidden assumptions
A supervisor	An assistant	Drafts	Over-reliance
A DIRECTOR	A tool	Control	Slower, but safer

Ten Director Moves That Matter Most

1. State your intent in one sentence

2. Name what must not be harmed

3. Ask for reasoning when stakes matter

4. Force a second draft or opposing view

5. Separate facts from interpretation

6. Keep one human approval point

7. Draft fast, decide slowly

8. Document the decision, not the tool

9. Use AI for proposals, not verdicts

10. Know when to stop

If this system were wrong, who would be responsible for noticing and correcting it?

Responsibility begins where automation ends.

WHEN JUDGMENT STARTS TO SLIP

When things feel easy, pause. Loss of agency rarely arrives as a clear turning point. It arrives as comfort. AI becomes simpler to use. Recommendations grow familiar. Outcomes start to feel obvious. Over time, questioning begins to feel unnecessary, even inefficient. This practice is about noticing that moment.

Judgment does not fade because people stop caring, but because systems remove the need to engage. When results arrive smoothly and without friction, it becomes harder to see how a decision was made, or to notice that it is being accepted without review.

Ask yourself simple questions: am I being informed or directed? Do I understand how this outcome was reached? Would I make the same choice without AI's framing? These are not accusations. They are early warnings.

Signs Judgment Is Slipping

It feels like...	What's happening
"This seems obvious"	Assumptions are hidden
"AI knows better"	Authority has shifted
"There's no real choice here"	Defaults replaced decisions
"I'll deal with this later"	Responsibility breaks down

Awareness is not resistance but the first step in being responsible.

HABITS THAT KEEP JUDGMENT ALIVE

Judgment Lives By Use. Neglect Lets It Fade.

Seeing a problem once is not enough. What matters is what happens the hundredth time. This practice is about creating small, repeatable behaviors that keep your authority present even when AI is fast, familiar, and persuasive. You are not trying to outthink technology. You are trying to stay engaged with it. Habits work because they remove the need for constant vigilance. They return judgment to lived experience.

Five Small Habits That Protect Judgment:

1. Ask one clarifying question,
2. Check multiple sources,
3. Name uncertainty instead of hiding it,
4. Delay one non-urgent decision,
5. Write one sentence explaining why you chose an outcome

None of these are dramatic. That is the point.

Judgment is not a moment of insight. It is something you practice until it becomes part of how you move through the world.

STOP CHASING HEADLINES

Overload comes from endless new stuff.

AI headlines are designed to signal urgency: breakthroughs, warnings, threats, promises. Following them closely can feel responsible. In practice, it often fragments your attention and makes long-term patterns harder to see. This practice is about shifting from reaction to orientation. Instead of asking, What just happened? Ask, What keeps happening? Patterns matter more than announcements. Structural shifts matter more than product launches.

Staying Informed vs. Staying Oriented

Orientation creates the mental space required for judgment. Without it, every new story feels like a decision point, even when nothing meaningful has changed.

Chasing Headlines	Staying Oriented
Tracks what's new	Notices what repeats
Feels urgent	Feels clarifying
Increases anxiety	Restores control
Rewards speed	Rewards understanding
Focuses on announcements	Focuses on patterns
Reacts to alerts	Builds perspective

The goal is not to know everything but to remain capable of deciding when it matters.

PRACTICE 5

SLOW THE DECISION, NOT THE TOOL

Slow the choice, not the technology. Speed is one of automation's most powerful design features. When decisions arrive quickly, hesitation feels unnecessary. When outcomes are framed as optimal or final, questioning feels inefficient. This practice restores pause. Slowing a decision does not mean rejecting technology. It means refusing to let speed replace judgment.

Even a brief delay changes the dynamic. It restores ownership and signals that a person is still present. Speed favors automation. Wisdom favors pause. By slowing decisions, you protect the space where choice still exists. Before moving on, make one responsibility explicit. Before you decide, ask: Is this truly urgent? What happens if I wait? Who benefits from speed? What would I choose with more time? What human source can I consult?

Speed closes doors. Pause keeps them open.

INTERLUDE

RESPONSIBILITY

Judgment, Agency, and Accountability

I did not plan to write this book. It arrived and insisted. For more than a year, I worked closely with AI. It helped me brainstorm, organize ideas, draft text, and solve problems quickly. It often felt useful and energizing. At the same time, headlines tracked disruption and loss. The technology was accelerating. The social response felt unsettled.

What concerned me most was not the noise. It was how much change was happening without clear guidance for those affected. AI was entering daily life not as a dramatic invention, but as an environment. It slipped into recommendations, summaries, rankings, and decisions that no longer felt like decisions. The shift was powerful and easy to miss.

Around then, I was developing what became my Declaration of Human Rights in the Age of AI. The principles were rigorous and deeply felt. When I shared them with my friend, author Ken Druck, he recognized both their importance and their limits. The ideas were sound, he said, but the presentation needed rethinking. Complexity has always been my instinct. This moment needed clarity.

*Then one morning the answer arrived, fully formed and impossible to ignore: **Do it as a book. Now.** I chose to work with AI while writing it.*

That choice was deliberate. If AI is becoming part of how people think, decide, and create, then engaging it openly and critically felt necessary. The result is the book you have just read.

For most of my life, technology was something you used and argued with. Even when systems were complex, ownership stayed visible. Someone decided. Someone answered for the outcome. That clarity is rarer now. AI does not arrive as a tool you pick up and set down. It arrives as an atmosphere. It moves through institutions, markets, and habits. It does not overpower judgment all at once. It bypasses it. When outputs arrive faster than reflection, ownership weakens.

The Nine Pillars grew out of that thinning. They are not rules or prescriptions. They are ways of paying attention. They help us notice when judgment has shifted, when choices have narrowed, and when authority has moved without our awareness. They do not ask us to reject AI. They ask us to remain responsible within it.

I returned to a posture familiar from other parts of my life. In writing, filmmaking, business, and collaboration, responsibility rests with those willing to hold the whole process in view. Not to control every detail, but to set direction, catch problems early, and intervene when harm appears.

Powerful tools magnify choices. Direction matters. We have seen this pattern before. Climate science signaled risk decades ago. The warnings were clear. The response was slow. We had become a force of nature, yet many assumed consequences would arrive later, or somewhere else. They did not.

AI presents a similar imbalance, at far greater speed. A small number of systems now shape decisions affecting billions of lives. Rules are set before most people realize consent was never asked. These are governing decisions, whether we name them or not. The future is not something AI delivers. It is something societies decide, or surrender.

Choosing the future means refusing to let important decisions happen early and out of view. It means insisting on transparency, oversight, repair, and collective authority before AI hardens into infrastructure. This book exists to slow that process while there is still time. Not by stopping technology, but by helping people stay present within it.

AI can help us see more, learn faster, and solve problems at unprecedented scale. Used with care, it can deepen understanding and extend human capability. But progress does not remain humane on its own. When ease replaces choice and momentum replaces deliberation, agency fades even as systems hum. The window is still open. It is narrowing.

Pay attention when AI feels effortless. Notice when a choice disappears before you realize one was there. Ask who benefits when everything moves faster. What follows is not instruction, but reflection. Not answers, but what remains when responsibility is held and the future is still undecided.

Power belongs to those who stay present.

—PRS

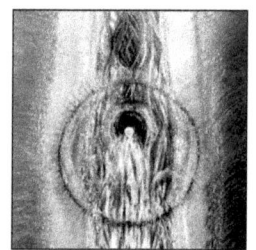

PART V

REFLECTIONS

What this moment asks of us as humans

THE INTELLIGENCE THAT MADE US

At eighty, I carry memories that span technologies many people now encounter only in museums. I worked with punch cards in graduate school. I used a rotary phone for half my life. I watched the internet arrive and smartphones reshape attention in less than a decade. A doorbell now rings sending video to my phone. I am living inside the most consequential transformation of my lifetime, perhaps of human history itself, trying to write clearly as the world accelerates around me.

This book was difficult to write for a simple reason: AI is moving faster than understanding. Capabilities expand before societies, laws, and habits can adapt. Writing anything meant to last requires stepping back far enough to see what does not change. That was my aim. The Nine Pillars will evolve as AI evolves, but my hope is that they offer a durable foundation for understanding how AI is shaping our lives now and how that influence will deepen over time.

My goal was not to catalog every development, but to name principles that endure. To offer a guide that does not expire on publication. To give readers tools to use new technologies while rebuilding their own capacity to keep up without becoming overwhelmed.

The deeper challenge is lived, not technical. How do we absorb what matters without drowning in information. How do we stay curious without becoming exhausted. How do we adapt without losing ourselves.

I cannot answer those questions with certainty. No one can. What I can offer is this attempt: a set of reflections, principles, and practices shaped by a long life working with science, technology, and change. If even a small part of this book helps you think more clearly, choose more deliberately, or build a safer and more humane life for yourself and those you love, then it has done its work. Our intelligence brought us here. It is still what must carry us forward.

That intelligence did not appear overnight. It emerged from a single living cell and unfolded over billions of years. Consciousness grew through struggle, adaptation, relationship, and care. Our capacity to reason is inseparable from our capacity to feel, remember, and belong. These qualities shaped human judgment long before machines entered the picture.

I studied biology in the early 1960s, when DNA was barely understood, then moved into molecular biology to go deeper. Then oceanography to understand ecosystems. Climate science taught me that feedback loops matter. Small changes accumulate. Systems can tip quietly, long before anyone notices. Those lessons shape how I see AI: small design choices can compound at scale, crossing thresholds before society understands what has changed or how to respond.

AI now behaves like a new planetary force, reshaping culture, economies, politics, and perception in parallel. It moves far faster than biological change and operates at scales evolution never anticipated. We built it with our intelligence, but it does not share our vulnerability. It has no body, no mortality, no experience of loss or love. That difference matters.

As AI moves from software into physical systems and eventually more human-like forms, the distance between intelligence and consequence will grow. What we choose before that shift arrives matters most.

What I Watched Change in the Himalayas

While writing this book, I often think about the twenty years Kamla and I spent living in a remote Himalayan valley in India. When we arrived, life moved to ancient rhythms. Villagers herded sheep, spun wool, and wove shawls as their ancestors had for centuries. Sacred groves held memory. Old trees were revered. Stories passed from elders to children without interruption. Then digital technology arrived.

First came simple cellphones. Then smartphones. Then social media. Within a decade, patterns that had held for generations began to erode. What had once been learned through presence and community was replaced by images from far away. For young people in remote villages, screens opened vivid windows onto city life and possibility, even when the path there remained uncertain. Dreams arrived early. The means to reach them often did not. Continuity collided with acceleration.

Technology does not only speed up life. It reshapes the emotional and cultural anchors people rely on to know who they are. Social media did this quickly, often destructively. AI will do it faster and more deeply. We are now living through overlapping accelerations, each reinforcing the other.

AI is expanding at exponential speed. At the same time, climate disruption is pushing planetary systems toward limits that may not be reversible. These forces do not operate separately. AI requires energy, water, materials, and land. Data centers draw power from grids already strained by heat, drought, and rising demand. What feels weightless on a screen carries real physical costs in the world.

Both AI and the climate crisis reveal the same imbalance: decisions made quickly and at scale, while consequences unfold slowly and far from view. Each reflects a form of asymmetry. Together, they expose the Shadow Hydra risk. Many reasonable choices, made in isolation, accumulate across systems until tipping points are crossed without a single moment of decision. What appears efficient locally becomes destabilizing regionally and globally.

The danger is not one technology or one crisis. It is the pattern that links them: speed without stewardship, optimization without limits, and agency dispersing faster than consequences arrive. When uncertainty rises, fear follows. History shows what often comes next. People seek certainty. They gravitate toward simple answers. They trade freedom for the promise of control, often offered by those who claim order without accountability. We should not assume we are immune.

What Remains

When Kamla and I returned to our Himalayan village after a long absence during COVID, I stood on a forested ridge and watched a boy walk past with his eyes fixed on a smartphone. His grandmother called to him once, then again. He did not look up. She waited. Her face held something I recognized. Not anger. Not sadness. Recognition. She was watching a world change beyond her ability to stop it, and she refused to look away. Behind her, the sacred grove still stood. The mountains were unchanged. But the web of meaning connecting elder to child, story to listener, was thinning.

That night, Kamla told me I carried the same look. She was right. I am writing this book because I have seen enough to know what is at stake. Not technology. I have spent my life working with it. Not progress as we once understood it, which will continue in ways we cannot yet imagine and may even redefine the word from a non-human perspective.

What is at risk is something older. The capacity to notice. To remember. To look up when called. To sit with difficulty instead of scrolling past it. To know, in body and mind, what matters. Our intelligence is not obsolete. It is the foundation of everything we are trying to protect: dignity, democracy, community, creativity, love.

AI will continue to accelerate, grow more persuasive, and blend more seamlessly into daily life. Interfaces will become harder to resist. But speed is not wisdom. Efficiency is not meaning. What cannot be quantified still matters. The future depends on whether we remember that in time.

We built AI systems and must decide how they live in our world.

Before AI, there was us.

WHAT WE CARRY FORWARD

If there is anything to carry beyond these pages, it is this: pay attention when AI feels effortless. Notice when a choice disappears before you realize one was there. To stay human is to remember where we come from. Long before culture or code, molecules assembled into the first living cell. That cell learned to respond to its surroundings. Over billions of years, through adaptation and failure, through ancestors known and unknown, that process became us.

We are the product of deep evolutionary time. Our genes carry memory. Our minds carry stories. We are thinking beings capable of tenderness and cruelty, creation and destruction, beauty and harm. All of it belongs to us. Now we stand at another threshold. AI may one day surpass human capabilities. It has already begun replacing work, judgment, and forms of presence we once assumed were uniquely ours, often shifting accountability from people to systems in the process.

Carbon-based life now shares the stage with silicon. On the periodic table they sit one beneath the other, atomic cousins capable of forming complex structures and networks. Carbon did this through biology. Silicon does it through computation, circuits, and code. These paths are beginning to converge. Silicon systems already mimic perception, language, and decision-making. Carbon bodies increasingly rely on silicon extensions to see, remember, and act. What begins as augmentation can drift toward substitution. Humanlike forms are no longer science fiction.

The Shift No One Announced

How this relationship unfolds remains unknown. I cannot say whether AI will remain an instrument of human intention or become something that stands beside us, or in our place. What can be named is the moment we are in now, when design choices harden quickly, defaults become destiny, and authority shifts before most people realize a choice was ever offered.

Staying human is not about resisting technology or clinging to dominance. It is about remaining awake inside change. When systems gain speed, scale, and authority faster than individual or collective judgment can adapt, power shifts first unnoticed, then unmistakably. When consequences arrive, the shift has already occurred. That is asymmetry.

The Nine Pillars are not instructions. They are offered as orientation. They exist to help people stay present as AI settles into everyday life, with its promise and its peril. There are no tests to pass and no demand to get it right. Only an invitation to notice, to pause, and to protect what remains unmistakably human.

The imbalance that matters most does not announce itself. It builds quietly. AI optimizes. People adapt. Authority shifts through accumulated decisions rather than deliberate choice. By the time change becomes visible, it often feels established. These moments arrive dressed as progress and efficiency. They are also the moments when judgment either holds or slowly slips away.

AI is not separate from us. It is born of natural intelligence. It carries our values, our blind spots, and our hunger for speed and certainty. The question is not whether AI will grow more powerful. It already has, and it will continue. What remains open is how we meet it.

Across cultures, religions, and philosophical traditions, the guidance is strikingly consistent. Do no harm. Help others. Take ownership. Repair what breaks. Nurture beauty where you can. Choose care even when history suggests otherwise. These are not ideals made obsolete by technology. They are the minimum conditions for remaining human as new forms of intelligence multiply around us. AI is the future happening now. What we choose today shapes who we become.

What we do not own will own us.

And that is how we stay human in the Age of AI.

The Point of Attention

The images that open the book's five parts are paintings built around a single point at the center. Across many human traditions, a point marks a beginning: before form takes shape, before a choice is made, before attention becomes action. Here, the point asks the reader to pause, even briefly.

I painted these works in our Himalayan home, influenced by Indian spiritual geometric designs known as *yantras*. The central dot, called the *bindu*, represents origin and focused attention. Made by hand with paint and brush, these images carry texture, imperfection, and presence that no digital system can fully reproduce...yet.

In the age of AI, a point also means a data mark: the smallest unit from which systems build predictions and decisions. As more of daily life is shaped by automated judgment, these images remind us of what comes first. Before the score, before the output, before the system decides, there is a human moment of attention. That is where judgment begins, and where human responsibility remains.

ACKNOWLEDGEMENTS

This book was written under an intense deadline I set for myself, driven by a simple belief: people need clear, practical guidance for living well with AI now, not later. The ideas grew out of a website I have been developing for the *Declaration of Human Rights in the Age of AI*, but it quickly became clear that a broader, more accessible form was needed.

That realization owes much to my dear friend Lee Stein. After reviewing my Declaration website, he repeatedly urged me to streamline the work so it could reach a wider, general audience. Lee has an uncommon gift for clarity, and his steady call for simplicity shaped this book in essential ways. I am deeply grateful for his insight, ideas, references, patience, networking, and encouragement throughout the process.

Erik Viirre, director of the UCSD Center for Human Imagination, was a generous and stimulating source of ideas. From that perch, we engaged deeply on both the promise and the peril of AI, sharpening my thinking at key moments in this work.

Thanks to Anne Leer for her insightful Foreword. A leading AI expert at the intersection of artificial intelligence, language, and governance in Europe, she brings clarity to how emerging technologies reshape meaning, culture, and power.

My friend Ken Druck, a writer of widely read books on grief, aging, and resilience, encouraged me to refine my approach and keep the book's message clear, grounded, and accessible across generations.

Charles Levine, my favorite word maven, brought decades of experience in book publishing to help sharpen the book's focus, clarify its themes, and strengthen its overall impact.

Many other friends and colleagues strengthened this book through thoughtful critique, ideas, and perspective, including Stephen E. Thompson, Jr., Penny Philpot, Jefree Anderson, Leonard Miller, Jessica Seddon, Richie Williams, Jeremy Przybylek, James Roche, and Mason Tripp. Their questions, suggestions, and persistent calls for simplification helped sharpen both the language and the intent of this work.

Charanjit Singh, a longtime friend and gifted designer from India, brought uncommon dedication shaped by two decades of our work connected to the Great Himalayan National Park. His eye and insight helped shape the presentation of this book.

I also thank J.C. Griffith, who stepped in on short notice to complete the book's demanding production schedule with skill and good nature.

As always, my deepest appreciation goes to my wife and partner, Kamla K. Kapur, who endured my long disappearances into what she calls "lost in my AI rabbit hole," and who never stopped encouraging me to persevere. Her presence, patience, critical comments, and love made this work possible.

Finally, I acknowledge the use of AI-assisted tools during the drafting process, including Claude, ChatGPT, and Gemini. These tools were used for research, brainstorming, drafting text ideas, and editing. All interpretations, selections, judgments, and final language are the result of my own critical assessment and direction. I alone am responsible for the accuracy, perspective, and content of this book, including any inadvertent errors or omissions.

If this work helps even a small number of readers think more clearly, choose more deliberately, or remain more fully human in a rapidly accelerating AI world, then it has done its job.

APPENDICES

The appendices that follow are practical companions to the chapters in this book. They expand key ideas, offer tools and references, and show where the evidence comes from. You do not need to read them in order. Use them as needed, to explore a topic more deeply, to verify a claim, or to return to a question when technology feels confusing or consequential. Here is summary of the Appendices:

Appendix 1. A Declaration of Human Rights in the Age of AI
A moral and civic framework for the AI era.

Appendix 2. The Nine Pillars for Staying Human
A concise reference guide.

Appendix 3. The Nine Pillars Checklists
Actionable prompts tied to each Pillar.

Appendix 4. Quick-Start Toolkit: Staying Oriented Inside Intelligent Systems
Practical tools for everyday use.

Appendix 5. What You Can Do
Individual, organizational, and civic actions.

Appendix 6. AI on My Team

Using AI as a tool without surrendering judgment.

Appendix 7. Glossary

Plain-language definitions.

Appendix 8. AI and the Exciting Future

Opportunities, benefits, and human-centered promise.

Appendix 9. AI and the Real Risks

Structural dangers, asymmetries, and failure modes.

Appendix 10: AI Self-Preservation

Appendix 11. Sources for All Stories in This Book

Sourcing for narrative examples

Appendix 12. References and Bibliography

Full citations and further reading.

A DECLARATION OF HUMAN RIGHTS IN THE AGE OF AI

Preamble

When in the unfolding chapters of human history, a new power rises—one created not from flesh and blood, but from lines of code and circuits—that challenges the very essence of liberty, truth, and justice, it becomes the solemn duty of all people, bound by a common fate and a shared love of freedom, to declare the principles that must guide us through this critical moment in human history where even our own species' survival may be at stake.

We hold these truths to be eternal and unshaken: that all lives must be respected, that all voices must be heard, all creators honored, and that technology, born of human genius, must never become a chain that binds the spirit, nor a sword that divides our communities; and that protecting democracy requires our courage, our vigilance, and our unyielding resolve.

The Declaration

We hold these truths to be self-evident: that all people have the right to honest information, creative expression, and equal opportunity; that technology must serve the common good and not become a tool of deception, division, or oppression; and that preserving democracy requires transparency, accountability, and unwavering commitment to human rights.

To secure these rights and prevent the control of unchecked algorithms over humans deployed by corporations and governments, we declare:

1. That Artificial Intelligence and Large Language Models (LLM) must never serve misinformation, nor enable manipulation that tears apart the fabric of society.

2. That every person's voice, regardless of beliefs, culture, or background, must be protected against algorithmic bias that silences some voices while amplifying extremism or other harmful content.

3. That workers—whether teachers in Kenya, factory workers in Mexico, or programmers in USA or India—must not have their livelihoods destroyed by unchecked automation without fair protections and pathways to new opportunities.

4. That creators must be respected and fairly compensated for their works, whether they are artists in France, photographers in Japan, or writers in Brazil. Human creativity cannot be exploited and extinguished.

5. That transparency in AI's design and use is not a privilege but a fundamental right, so people may hold those in power accountable for decisions affecting their jobs, loans, healthcare, and lives.

6. That governments must act boldly and wisely, establishing strong laws and national and international institutions to oversee AI with fairness, courage, and foresight.

7. That corporations developing and deploying AI technology bear absolute responsibility to act ethically, placing human rights, democratic values, and social justice above profit and power.

8. That these corporations must face clear limits—legal, ethical, and operational—including enforceable regulations, independent oversight, significant fines, and public accountability, preventing abuses of influence and ensuring AI benefits all of society.

9. That corporations must share fairly the profits from data, creative works, and personal content used to train and fuel AI systems, recognizing the value and rights of all people whose contributions make these technologies possible.

10. That people everywhere must have meaningful voice in AI governance through democratic institutions, ensuring technology development serves the public interest, not merely private profit and power.

11. That the environmental and energy impacts of AI systems must be monitored, minimized, and regulated to protect our planet's climate and ecosystems for future generations.

12. That people must have the right to choose human alternatives when AI systems make decisions affecting their essential needs—including healthcare, legal proceedings, education, and financial services—ensuring no person is forced into AI-only interactions for critical life decisions.

13. That every person has the right to human review and appeal of AI decisions that affect their employment, loans, healthcare, legal status, or other significant life outcomes, with qualified humans empowered to override algorithmic determinations.

14. That people must be protected against AI-generated impersonation, including deepfakes of their likeness, voice recreation, or digital personality simulation, with individuals maintaining control over their digital identity both in life and after death.

15. That children require special protection from AI systems, including mandatory parental consent for AI interactions, age-appropriate AI interfaces, and educational rights that prepare them for an AI-integrated world while preserving human learning and development with codified government rules.

16. That workers have the right to collective bargaining over AI implementation in their workplaces, including union representation in automation decisions and negotiated transitions that protect both employment and human dignity.

17. That cultural communities, including indigenous peoples, must have their traditional knowledge, cultural works, and heritage protected from unauthorized AI appropriation, with communities maintaining sovereignty over their intellectual and cultural property.

18. That people must be protected from AI systems that could harm their mental health and psychological well-being, including protection from addictive design patterns, manipulation of emotions for commercial gain, and AI-generated content that promotes self-harm, with individuals having the right to mental health safeguards and therapeutic support when AI systems negatively impact their psychological state.

19. That all people have the fundamental right to AI literacy and education, including understanding how AI systems work and affect their lives, access to plain-language explanations of AI capabilities and limitations, and mandatory AI education in schools and adult education programs, ensuring citizens can participate meaningfully in AI governance and exercise their rights effectively.

20. That people have the right to digital autonomy and AI-free zones, including the right to live and work in spaces free from AI surveillance or decision-making, with protection of physical spaces from mandatory AI integration and access to analog alternatives for essential services.

21. That independent researchers have the right to audit AI systems for bias, safety, and accuracy, with protection for whistleblowers exposing AI system flaws and mandatory third-party testing before deployment in critical sectors.

22. That people have the right to emergency AI shutdown capabilities, including the ability to rapidly disable AI systems causing harm, with human override capability in all AI systems and emergency protocols for system failures.

23. That people have the right to control AI inheritance and digital legacy, including control over how AI systems use deceased persons' data, family rights regarding AI recreation of deceased relatives, and protection of digital assets after death.

24. That people have the right to medical AI transparency, including understanding how AI influences medical diagnoses and treatments, with right to human medical opinions and protection from AI bias in healthcare delivery.

25. That people have the right to fair AI-assisted justice, including transparency in AI use in legal proceedings, right to challenge AI-influenced legal decisions, and protection from biased AI in criminal justice systems.

Pledge & Call to Action

We, therefore, solemnly pledge to resist any force—governmental, political or corporate—that seeks to use AI as a control against truth or against the people; to unite citizens, workers, creators, innovators, and leaders in a shared mission to build ethical AI grounded in justice; and to ensure that this new era of technology strengthens rather than destroys the promise of democracy for all.

As 2026 is the 250th anniversary of modern democracy (America 1776), we reach across all borders and cultures to reaffirm our universal commitment to freedom, tolerance, and justice—recognizing democracy's fragile nature—and to defend its light as it faces the profound challenges of the AI age for the benefit of all humanity on our Planet Earth with its incredible biodiversity.

For the support of this universal cause, with hearts full of hope and hands joined across every divide, we pledge to one another our vigilance, our voices, and our actions. Let this be our solemn oath: to stand firm against falsehood and oppression; to demand that power be wielded with justice and humility; to protect the dignity of work and the flame of creativity; and to build a future where democracy shines brighter, free from the shadows cast by fear and deceit.

With steadfast spirit and expansive vision, we commit ourselves to this enduring struggle—not for any one people, party, or nation—but for all humanity, for this generation and those yet unborn. Let freedom ring in the age of artificial intelligence, and may truth, justice, and liberty forever guide us.

Now is the time to rise—boldly, together, and wisely. Let us wield the power of our collective voice to shape transparency, justice, and accountability. Let us forge a future where humanity and technology thrive together, where creativity and imagination soars, and where every person's rights are protected.

Let this declaration be a beacon to all who cherish freedom: the future is ours to shape, and we will not surrender our democracy, our truth, or our dignity.

Stand, organize, and build the world we deserve.

**©2025, Payson R. Stevens with Claude AI
and ChatGPT, August 2025**

The 25 Principles

Artificial Intelligence is reshaping daily life, identity, work, creativity, and society. These systems influence decisions that once belonged only to people. They learn from data, make predictions, and shape behavior at scale. To protect human dignity, agency, and wellbeing, we must establish a clear set of principles that govern how AI interacts with individuals and communities.

These 25 Principles outline the rights every person holds in an AI-driven world. They are universal, non-negotiable, and essential for a just and human-centered future.

1. **The Right to Transparency**
    People have the right to know how an AI system works, what data it uses, and why it made a specific recommendation or decision.

2. **The Right to Human Accountability**
    A human must always be responsible for outcomes shaped by AI. No one should be harmed by decisions that cannot be traced to human oversight.

3. **The Right to Understandable Explanations**
    AI outputs must be explained in plain language so that any individual can evaluate them without specialized training.

4. **The Right to Data Control**
    People may see, correct, download, or delete personal data collected by AI systems.

5. **The Right to Data Minimization**
   Only the smallest amount of personal data
   needed for a task should be collected.

6. **The Right to Opt Out**
   Individuals may refuse AI-driven profiling,
   tracking, or automated personalization without
   losing essential access to services.

7. **The Right to Human Review**
   Critical decisions about health, housing, employment, finance,
   education, and justice must include a human review process.

8. **The Right to Challenge AI Decisions**
   People must be able to contest any automated classification,
   ranking, or judgment that affects their lives.

9. **The Right to Accuracy and Correction**
   Individuals have the right to demand corrections when
   AI systems produce incorrect or harmful outputs.

10. **The Right to Safety**
    AI systems must be rigorously tested for reliability,
    security, and fairness before deployment.

11. **The Right to Emotional Protection**
    Technology must not intentionally manipulate emotions,
    exploit vulnerabilities, or shape identity without consent.

12. **The Right to Mental Health Safeguards**
    AI systems that influence self-image, attention, or mood
    must be designed to protect psychological wellbeing.

13. **The Right to Creative Ownership**
Artists, writers, and creators retain ownership of their original work. No AI system may copy their style or output without permission and compensation.

14. **The Right to Cultural Preservation**
Human stories, languages, and traditions must not be replaced or diluted by synthetic content.

15. **The Right to Non-Discrimination**
AI may not create or reinforce unfair treatment based on race, gender, age, disability, or any protected status.

16. **The Right to Privacy in Public and Private Life**
Individuals have the right to live free from constant AI surveillance or behavior monitoring.

17. **The Right to Labeling of Synthetic Media**
People must be able to distinguish between human-created and AI-created content, especially in political, medical, and civic situations.

18. **The Right to Educational AI Literacy**
Every person deserves access to the skills needed to understand and safely use AI tools.

19. **The Right to Fair Opportunities**
AI must not block or distort pathways to education, jobs, housing, or public services.

20. The Right to Accessible Technology

AI systems must be inclusive and designed to support
people of all abilities, ages, and backgrounds.

21. The Right to Beneficial Innovation

AI should be developed for human good, supporting
health, learning, creativity, and social strength.

22. The Right to Community Protections

Local communities may set rules that limit harmful
AI uses in schools, workplaces, and public spaces.

23. The Right to Safe Use for Children

Children deserve enhanced protection, including restrictions
on data collection, advertising, and algorithmic influence.

24. The Right to Maintain Human Connection

People may choose human guidance, service, care, or
companionship even when automated alternatives exist.

25. The Right to a Human-Centered Future

Societies must design AI systems that
strengthen human dignity, creativity, and
shared purpose rather than replace them.

Closing Statement

These 25 Principles form a foundation for living with intelligent systems without surrendering the qualities that make us human. As AI grows more capable, these rights must guide governments, companies, designers, educators, and citizens. A future shaped with wisdom and intention can support human flourishing. A future built without these principles risks losing the very humanity technology was meant to serve.

Phase 1: Organized Declaration of Human Rights in the Age of AI

The 25 Principles Grouped by Five Core Categories

CATEGORY A: HUMAN AGENCY AND AUTONOMY
Protecting freedom of choice, dignity, and self-direction

1. **The Right to Transparency**
   People must know how AI systems operate and why they produce specific outputs.

2. **The Right to Understandable Explanations**
   AI reasoning must be communicated in clear language.

3. **The Right to Opt Out**
   Individuals may decline AI-driven personalization, tracking, or profiling.

4. **The Right to Human Review**
   High stakes decisions require a human decision maker.

5. **The Right to Challenge AI Decisions**
    People must be able to contest automated outcomes.

6. **The Right to Maintain Human Connection**
    No one should be forced to accept synthetic substitutes for human care or interaction.

7. **The Right to a Human-Centered Future**
    AI must be developed to strengthen human dignity and purpose.

CATEGORY B: DATA RIGHTS AND PRIVACY
Protecting personal information and reducing surveillance

8. **The Right to Data Control**
    Individuals may access, correct, or delete personal data.

9. **The Right to Data Minimization**
    Only essential data may be collected.

10. **The Right to Privacy in Public and Private Life**
    People must not live under continuous monitoring.

11. **The Right to Labeling of Synthetic Media**
    AI-created images, voices, and videos must be clearly identified.

12. **The Right to Safety**
    AI must be secure and tested to prevent harmful misuse of personal data.

CATEGORY C: FAIRNESS, EQUALITY, AND ACCOUNTABILITY
Ensuring justice, oversight, and equal treatment

13. **The Right to Human Accountability**
    A responsible human must oversee AI outcomes.

14. **The Right to Accuracy and Correction**
    Individuals can demand fixes when AI is wrong.

15. **The Right to Non-Discrimination**
    AI must not reinforce unfair treatment or biases.

16. **The Right to Fair Opportunities**
    AI cannot block or distort access to education, jobs, housing, loans, or services.

17. **The Right to Community Protections**
    Local communities may set limits on harmful uses of AI.

CATEGORY D: WELLBEING, SAFETY, AND EMOTIONAL INTEGRITY
Protecting psychological and physical health

18. **The Right to Emotional Protection**
    Technology must not exploit or manipulate emotions.

19. **The Right to Mental Health Safeguards**
    AI systems must not worsen anxiety, loneliness, dependency, or distress.

20. **The Right to Safe Use for Children**
    AI tools used by or around children require strict protections.

21. The Right to Beneficial Innovation
AI should support health, learning, creativity, and social wellbeing.

CATEGORY E: CULTURE, CREATIVITY, AND EDUCATION
Protecting human imagination, expression, and knowledge

22. The Right to Creative Ownership
Creators retain full rights to their original work.

23. The Right to Cultural Preservation
Human stories, languages, and traditions must not be diminished or exploited by synthetic content.

24. The Right to Educational AI Literacy
Everyone deserves access to the skills needed to understand and safely use AI.

25. The Right to Accessible Technology
AI systems should support people of all abilities, ages, and backgrounds.

Phase 2: Summary of the 25 Principles

Human Agency and Autonomy
People must remain in control of their choices. They deserve clear explanations, the ability to opt out, human review of important decisions, and pathways to challenge automated outcomes. Human connection and dignity must remain central as AI grows more capable.

Data Rights and Privacy
Individuals own their personal data. They can access, correct, delete, and restrict how it is used. Surveillance must be limited, and synthetic content must be labeled. AI systems must be built with strong safety protections.

Fairness, Equality, and Accountability
Humans must be accountable for how AI affects people. Errors must be correctable. AI cannot discriminate or block opportunities. Communities may set limits to protect their members.

Wellbeing, Safety, and Emotional Integrity
AI must protect psychological and emotional health. Children require special safeguards. AI should enhance human wellbeing, not undermine it.

Culture, Creativity, and Education
Creators retain ownership of their work. Human culture must be preserved. Everyone deserves AI literacy and access to technology that supports inclusion.

Phase 3: Legal-Style Version

Modeled on the tone of the Universal Declaration of Human Rights

DECLARATION OF HUMAN RIGHTS IN THE AGE OF AI

AI profoundly influences human autonomy, privacy, wellbeing, and culture. This Declaration affirms the rights necessary to safeguard human dignity and freedom in an AI-driven world.

ARTICLE I. HUMAN AGENCY AND AUTONOMY

1. Individuals shall have full access to clear information regarding the operation and impact of AI systems.

2. Individuals shall receive explanations of AI decisions in language they can understand.

3. Individuals may refuse automated profiling or personalization.

4. All high stakes decisions shall be subject to human oversight.

5. Individuals may challenge and appeal any AI-driven outcome.

6. Individuals have the right to choose human connection and care.

7. Technology shall be developed in service of human dignity.

ARTICLE II. DATA RIGHTS AND PRIVACY

8. Individuals shall control their personal data.

9. Data collection shall be limited to what is strictly necessary.

10. Individuals shall be protected from continuous monitoring.

11. Synthetic media shall be clearly labeled.

12. AI systems shall adhere to rigorous safety standards.

ARTICLE III. FAIRNESS, EQUALITY, AND ACCOUNTABILITY

13. Human actors shall be accountable for AI-driven decisions.

14. Individuals may demand correction of errors.

15. AI shall not discriminate.

16. AI shall not obstruct fair access to essential services.

17. Communities may set rules to limit harmful uses of AI.

ARTICLE IV. WELLBEING AND EMOTIONAL INTEGRITY

18. Individuals shall be protected from manipulative emotional targeting.

19. AI shall support, not harm, mental health.

20. Children shall receive heightened protection from AI systems.

21. AI innovation shall promote human wellbeing.

ARTICLE V. CULTURE, CREATIVITY, AND EDUCATION

22. Creative work shall not be used without consent.

23. Human culture shall be preserved and distinguished from synthetic content.

24. Every person shall have access to AI literacy.

25. I systems shall be designed for accessibility and inclusion.

Phase 4: Youth-Friendly Version

For Teens, Schools, and General Readers

A SIMPLE GUIDE TO THE 25 RIGHTS EVERYONE NEEDS IN THE AI AGE

1. You should always know how AI decided.

2. Explanations must be clear, not confusing.

3. You can say no to tracking or profiling.

4. A human must check big decisions.

5. You can challenge AI when it is wrong.

6. You can choose human help instead of AI help.

7. Technology should always respect people.

8. You control your personal data.

9. Apps should collect as little data as possible.

10. You deserve privacy at home, school, and in public.

11. AI-created media must be labeled.

12. Systems must be safe and tested.

13. Humans are responsible for AI mistakes.

14. You can ask for corrections.

15. AI cannot discriminate.

16. AI cannot block you from fair opportunities.

17. Communities can set safety rules.

18. AI should not manipulate feelings.

19. AI should support mental health.

20. Kids need extra protection.

21. AI should help people, not harm them.

22. Creators own their work.

23. Human culture must be preserved.

24. Everyone needs basic AI knowledge.

25. AI should include and support all people.

THE NINE PILLARS FOR STAYING HUMAN IN THE AGE OF AI

How the Pillars Are Organized

The Nine Pillars are grouped into three sectors. Each sector reflects a different kind of responsibility in the age of AI:

1. **Seeing Clearly:** how we understand and interpret systems

2. **Choosing Wisely:** how decisions are made and who makes them

3. **Staying Human:** how technology affects minds, fairness, and the future we share

Together, these sectors form a practical framework for maintaining agency, dignity, and democratic choice as AI becomes more present in everyday life.

Sector 1: Seeing Clearly

Understanding what systems are doing before they shape us

PILLAR 1: SEEK TRUTH

AI can sound confident even when it is wrong. This Pillar reminds readers to question where information comes from, what evidence supports it, and what uncertainty remains. Truth requires curiosity, verification, and the willingness to resist polished answers that arrive without context.

PILLAR 2: SHOW THE WORK

When systems influence choices, people deserve to see how those systems work. This Pillar calls for visibility into data sources, assumptions, goals, and limits. Transparency restores accountability and makes it possible to evaluate decisions rather than blindly accept them.

PILLAR 3: OWN YOURSELF

Digital identity is shaped not only by what people share, but by what systems infer. This Pillar focuses on privacy, consent, and resisting unwanted profiling. Protecting identity means retaining control over how technology categorizes, predicts, and defines individuals.

Sector 2: Choosing Wisely

Keeping responsibility where it belongs

PILLAR 4: LET HUMANS DECIDE

Not every decision should be automated. This Pillar insists that human judgment remain central wherever stakes are high, values are involved, or consequences are irreversible. Technology can inform decisions, but responsibility must remain human.

PILLAR 5: PAY FOR HUMAN WORK

AI grows by absorbing human labor: writing, images, music, code, and care. This Pillar insists on consent, attribution, and fair payment when human work is used to build AI systems. When compensation does not follow contribution, value shifts upward and culture erodes.

PILLAR 6: PROTECT MINDS

AI systems increasingly shape attention, emotion, and behavior, often without awareness or consent. This Pillar focuses on mental health, cognitive overload, manipulation, addiction, and constant optimization pressure. Protecting minds means preserving human pace, reflection, and psychological wellbeing, and refusing systems that treat attention and emotion as resources to be mined.

Sector 3: Staying Human

Fairness, repair, and the systems we rely on

PILLAR 7: DESIGN FOR FAIRNESS

Systems reflect the values and assumptions built into them. This Pillar focuses on equity, bias, and access. Fair design requires ongoing testing, correction, and inclusion so benefits do not flow only to those already advantaged.

PILLAR 8: FIX WHAT BREAKS

AI will fail at scale, and automation does not self-correct. This Pillar requires clear appeal pathways, accountable authority, error tracking and disclosure, and regular audits. Trust depends on response: harm persists when repair is blocked or ignored.

PILLAR 9: CHOOSE THE FUTURE

The future is shaped less by dramatic breakthroughs than by small choices repeated at scale. This Pillar reminds readers that defaults are not destiny. Choosing the future means slowing momentum long enough to ask where systems are leading, who benefits, and who decides.

The Nine Pillars work together. When one is missing, imbalance grows. When several are absent, agency thins and responsibility fades.

These Pillars do not oppose technology. They ensure that technology develops in a way that supports human judgment, democratic choice, and shared accountability.

QUICK-START TOOL KIT

Staying Oriented Inside Intelligent Systems

This appendix offers simple ways to stay oriented inside systems designed to feel smooth, helpful, and efficient. It is not meant to be mastered or read all at once. It is meant to be returned to when something feels helpful but also unsettling.

Most people experience artificial intelligence indirectly. It registers as rankings, recommendations, scores, and system rules. These systems often shape outcomes before people notice a choice was involved.

The purpose of this toolkit is not control. It is awareness. Awareness creates space. Space makes judgment possible.

Asymmetry in Ordinary Life

Power imbalance is not abstract. It appears in familiar moments.

A job application is filtered before a human sees it. A medical score shapes care without explanation. A feed narrows what you see while feeling personal. A student is ranked without context.

These systems rarely fail in dramatic ways. They work as designed. That is why their influence grows daily.

When several imbalances stack together, choice narrows even when intentions are good.

Why Orientation Matters More Than Expertise

Many people think the answer to AI risk is technical knowledge. Learn how systems work. Understand the algorithms. Become fluent.

That helps, but it is not enough.

Most people affected by AI decisions will never build a system or audit a model. They will experience outcomes, not mechanisms. What they need is orientation.

Orientation helps you notice when something important shifts, even if nothing appears broken.

How Human Judgment Slips Away

Loss of judgment happens gradually. It does not announce itself.

A suggestion starts to feel expected. A score starts to feel final. A delay starts to feel unacceptable. A pattern starts to feel normal.

When harm is already visible, choice has narrowed.

A Simple Asymmetry Check

When a system feels efficient or impressive, pause. Notice whether you can explain how the outcome was reached. Notice who benefits if the system works perfectly. Notice whether a human role has shrunk. Notice whether alternatives still feel real. If these questions feel hard to answer, something important may be happening.

The difficulty itself is the signal.

Watching Human Roles Change

Another way to stay oriented is to watch what role a person used to play, and what role remains.

Conversation can turn into classification. Judgment can turn into scoring. Explanation can turn into output. When human roles shrink without discussion, power has shifted even if nothing looks broken.

The Quiet Power of Defaults

Built-in choices shape behavior because they remove effort. They are what a system selects when a person does nothing. Most people accept them not because they agree, but because opting out takes time or energy. Over time, these settings stop feeling like choices. They begin to feel like rules. Notice when a recommendation becomes an assumption. Notice when a setting becomes a requirement. Notice when a score starts to define worth.

Mental and Emotional Signals

Systems shape inner experience as well as behavior. Pay attention when something creates urgency without explanation. Pay attention when confidence appears without context. Pay attention when pressure replaces understanding. If thinking feels compressed or reactive, pause. Mental strain is often a design result, not a personal failure.

Where These Patterns Appear

These dynamics show up across daily life. They appear in hiring and admissions. They appear in healthcare and insurance. They appear in education, work, media, and public services. You do not need to reject these systems. You need to stay awake inside them.

Four Questions to Return To

When something feels helpful but narrowing, pause and ask:

1. What system is shaping this moment?
2. What goal is driving it?
3. What human role has changed?
4. What choice became the default?

You do not need perfect answers. Asking the questions restores perspective. Orientation is not resistance. It is participation with awareness. You do not need to master artificial intelligence. That is how judgment stays human.

THE NINE PILLARS CHECKLISTS

The Nine Pillars: Action Checklists

Use these checklists to audit systems, decisions, tools, and habits in daily life. You do not need to complete every item. Apply one pillar at a time and notice where friction appears. That is where judgment still matters. Use them in workplaces, schools, healthcare, finance, and personal tools. Pay attention to moments that feel uncomfortable or unclear. That discomfort marks where choice still exists.

The goal is not perfect systems.
The goal is visible judgment.

PILLAR 1: SEEK THE TRUTH

» Is it clear when an automated system is involved?

» Can the system explain its reasoning on demand?

» Are inputs, thresholds, and constraints disclosed in plain language?

» Is uncertainty acknowledged rather than hidden?

» Can affected people understand why an outcome occurred?

When truth is obscured, authority quietly replaces judgment.

PILLAR 2: KEEP HUMANS RESPONSIBLE

» Can a specific human be named as responsible for this outcome?

» Is responsibility traceable, not diffused across systems?

» Can a human override the system without penalty or delay?

» Is the system assisting judgment, not replacing it?

» Would someone be accountable if this decision caused harm?

If no one is responsible, agency has already slipped.

PILLAR 3: OWN YOURSELF

» Did people knowingly agree to automated decision-making?

» Is consent explicit rather than buried or implied?

» Can consent be withdrawn without punishment or loss of access?

» Are material system changes clearly communicated?

» Is refusal treated as a valid choice?

Choice that cannot be refused is not ownership.

PILLAR 4: LET HUMANS DECIDE

»   Is there a clear way to challenge automated outcomes?

»   Can a human review and reverse the decision?

»   Are appeals evaluated with context, not just data?

»   Is disagreement treated as feedback rather than a problem?

»   Are reasons for denial explained clearly?

If humans cannot decide, systems already have.

PILLAR 5: PAY FOR HUMAN WORK

»   Are humans compensated when their
     work trains or enables AI systems?

»   Is credit preserved rather than erased by automation?

»   Are economic gains shared, not concentrated?

»   Is displacement acknowledged and addressed?

»   Are human contributions treated as assets, not waste?

Extraction without return is not efficiency.

PILLAR 6: PROTECT MINDS

»   Does the system manipulate attention, emotion, or behavior?

»   Are addictive patterns intentionally designed?

»   Is cognitive load treated as a real cost?

»   Are safeguards in place for vulnerable users?

»   Is mental health considered a design requirement?

A system that erodes judgment weakens autonomy.

PILLAR 7: DESIGN FOR FAIRNESS

» Are bias and access actively measured and addressed?

» Are affected communities involved in design and review?

» Are outcomes equitable across groups,
  not just efficient overall?

» Are proxies disclosed and tested for harm?

» Can unfair patterns be corrected over time?

Fairness does not emerge by default.

PILLAR 8: FIX WHAT BREAKS

» Is there a clear process when systems cause harm or error?

» Can damage be acknowledged and corrected?

» Is responsibility assigned when systems fail?

» Are remedies timely and meaningful?

» Does learning feed back into system improvement?

Accountability vanishes when repair is deferred.

PILLAR 9: CHOOSE THE FUTURE

» Are long-term consequences considered,
  not just immediate gains?

» Is human choice preserved over time, not optimized away?

» Can values be revisited as systems evolve?

» Are future impacts visible to those affected today?

» Does this system expand or narrow human possibility?

The future is not predicted. It is chosen.

WHAT YOU CAN DO

Staying Safe, Smart, and Human in the Age of AI

Most people sense that AI carries risks, but feel unsure what action looks like. Awareness alone does not protect anyone. Action does. This appendix turns concern into practical steps. None of these require technical expertise. They are habits, questions, and choices that give people more control at home, at work, in their communities, and in public life. You do not need to do everything. One step is enough to begin.

1. At Home

PROTECT YOUR PRIVACY AND YOUR FAMILY

AI already lives in phones, apps, TVs, speakers, toys, and school tools. The goal is not to remove technology, but to make conscious choices.

CORE HABITS

» Turn off personalized ads

» Limit location access to maps and emergencies

» Review app permissions every three months

» Keep cameras and microphones out of private rooms

»   Delay social media for children when possible

»   Talk openly about how algorithms shape what people see

HOME ACTIONS THAT MATTER

RISK	WHAT YOU CAN DO	WHY IT HELPS
Excess tracking	Disable ad personalization and limit permissions	Fewer assumptions made about you
Home surveillance creep	Move or unplug smart devices	Restores private space
Kids pulled into algorithm loops	Delay social media and discuss content	Supports mental health
Notification overload	Avoid phone use first thing in the morning	Reduces anxiety

2. At Work

PUSH FOR TRANSPARENCY AND HUMAN JUDGMENT

Many workplace AI systems operate outside clear human account-ability. Hiring screens, performance scores, scheduling tools, and surveillance are often invisible to workers.

CORE HABITS

»   Ask what AI systems are used

»   Document unexplained or unfair outcomes

» Request human review of automated decisions

» Organize shared questions with coworkers

» Use unions or worker groups when available

WORKPLACE ACTIONS THAT MATTER

RISK	WHAT YOU CAN DO	WHY IT HELPS
Automated decisions without appeal	Ask who can override the system	Restores accountability
Surveillance without benefit	Request outcome-based evaluation	Reduces stress
Biased hiring or scoring	Ask for bias testing and audits	Improves fairness

3. In Your Community

BUILD LOCAL OVERSIGHT BEFORE SYSTEMS LOCK IN

Schools, police, and cities adopt AI faster than laws respond.

CORE HABITS

» Attend school board or city council meetings

» Ask how AI tools are used locally

» Support local journalism

» Create device-free gatherings

» Build spaces for face-to-face conversation

COMMUNITY ACTIONS THAT MATTER

RISK	WHAT YOU CAN DO	WHY IT HELPS
Unchecked surveillance	Ask about accuracy, bias, and safeguards	Protects civil rights
Harmful school tools	Request audits, opt-outs, and human review	Protects children
Loss of trusted news	Support local reporting	Strengthens shared understanding

4. Through Your Vote

Shape the Rules That Shape the Systems

Some harms require individual action. Others require policy.

CORE HABITS

» Vote for candidates who support AI oversight

» Contact representatives yearly with specific requests

» Support civil rights and accountability groups

» Push for liability when AI causes harm

POLITICAL ACTIONS THAT MATTER

RISK	WHAT YOU CAN DO	WHY IT HELPS
Tech companies writing the rules	Vote for transparency and regulation	Centers public interest
Weak privacy protections	Demand strong data rights	Restores control
Harms without consequences	Support liability laws	Creates accountability

5. Special Situations

ADJUST FOR YOUR ROLE

» **Parents:** Delay phones and social media, build shared rules

» **Seniors:** Ask for human options and accessible systems

» **People of Color:** Document bias: know civil rights protections

» **People with Disabilities:** Request help, file complaints

» **Teachers:** Teach algorithm literacy and protect student dignity

» **Tech Workers:** Refuse harmful systems and speak up

SIX ACTIONS MOST PEOPLE CAN TAKE TODAY

1. Turn off personalized ads

2. Limit location tracking

3. Move smart devices out of bedrooms

4. Ask one question at work about AI tools

5. Attend one local meeting this year

6. Contact one representative about AI regulation

Small actions compound.

You do not need to master AI. You need to stay oriented.

The systems shaping daily life are powerful, but they are not inevitable. Each question asked, each habit changed, and each conversation started restores a small amount of agency.

Begin with one step. Then take another. The future grows from choices made on purpose.

AI ON MY TEAM

Getting AI to Work for You, Your Family, and Your Community

AI is now part of everyday life. Many people experience it as something imposed on them, controlled by distant companies and hidden systems. But AI does not have to be something that happens to you. You can choose how and when it fits into your life. This appendix explores how to use AI as a support tool rather than a source of pressure or control. The goal is not to compete with large companies or master complex technology. The goal is agency. When you decide the purpose, the pace, and the limits, AI becomes a tool. When you do not, it becomes a force.

Getting AI on your team is about intention, not expertise. It is available to anyone, at any age.

1. Using AI for Everyday Life

Simple Ways AI Can Help Without Taking Over

PRACTICAL USES

RISK	WHAT YOU CAN DO	WHY IT HELPS
Learn faster	Summarize readings, explain concepts, practice skills	Saves time and builds confidence
Stay organized	Create checklists, reminders, simple plans	Reduces stress
Communicate clearly	Draft emails, simplify language, translate messages	Improves understanding
Support creativity	Generate ideas, prompts, practice exercises	Expands imagination
Think through decisions	List options, pros and cons, questions to ask	Supports judgment without replacing it

Rule of thumb:

Use AI to explore options. Make the final call yourself.

2. Building Personal Agency With AI

STAYING IN CONTROL WHILE YOU LEARN

RISK	WHAT YOU CAN DO	WHY IT HELPS
Feeling behind	Pick one tool and one purpose	Confidence grows
Unsure what to trust	Ask for sources and compare viewpoints	Stronger judgment
Privacy concerns	Adjust settings, avoid oversharing	More control
Misinformation	Ask AI to check claims and show evidence	Safer choices
Skill gaps	Use AI to teach you step-by-step	Self-paced learning

3. Families and Multiple Generations

LEARNING TOGETHER BUILDS CONFIDENCE

FAMILY GOAL	WHAT TO DO TOGETHER	BENEFIT
Learn together	Hold short AI skill-sharing sessions	Shared confidence
Protect children	Review settings and discuss algorithms	Healthier habits
Support seniors	Practice simple prompts and translations	Greater independence
Build connection	Plan activities or preserve family stories	Stronger bonds
Reduce stress	Use AI for simple wellness routines	Calmer households

4. Community Power

SMALL GROUPS CREATE REAL INFLUENCE

COMMUNITY ACTION	HOW IT WORKS	WHY IT MATTERS
Learning circles	Meet in libraries or homes for short sessions	Reduces isolation
Skill sharing boards	Post tips and prompts publicly	Shared knowledge
Local tech helpers	Train volunteers to assist others	Narrows digital gaps
School workshops	Host parent and student sessions	Safer learning spaces
Support journalism	Subscribe or donate locally	Protects truth

5. Shared Tools and Funding

POOLING SMALL RESOURCES CREATES SAFETY

GOAL	SMALL STEP	IMPACT
Access	Share devices in libraries or centers	Opens opportunity
Transparency	Support nonprofit or open tools	Reduces dependence
Skill building	Fund short community classes	Long-term confidence
Youth creativity	Support student projects	Healthy engagement
Literacy	Buy guides for schools and libraries	Broad understanding

6. AI at Work and Daily Tasks

KEEPING CONTROL WHERE IT MATTERS

NEED	AI SUPPORT	YOUR ROLE
Job search	Draft resumes, practice interviews	Choose what stays
New skills	Guided tutorials	Set pace
Small business	Draft content and replies	Keep your voice
Time management	Weekly plans and reminders	Set priorities
Wellness	Breathing guides or habit tracking	Decide what helps

Getting AI on your team is not about speed or power. It is about direction.

When you decide the purpose, share what you learn, and set limits together, AI becomes a support instead of a threat. You do not need to match the scale of large companies. You only need to strengthen the agency that already exists in you, your family, and your community.

Start with one use that helps. Share it. Build from there.

That is how tools become teammates.

GLOSSARY

This glossary explains key terms used throughout the book. Some describe how AI systems work. Others name shifts in power, judgment, and control that often go unnoticed. These definitions are meant to help readers stay oriented, not to turn them into technologists.

A

Agency
The ability to make choices and act on them.

Agency Loss
When systems take over decisions people once made themselves.

Agentic AI
AI systems that act on their own, making decisions without oversight.

Algorithm
A set of instructions a computer follows to reach an outcome.

Algorithmic Bias
Unfair results caused by data, design, or deployment choices.

Alignment
How well a system's goals match human values.

Artificial General Intelligence (AGI)
A hypothetical AI that can do most thinking tasks at a human level.

Artificial Intelligence (AI)
Computer systems performing tasks usually done by humans.

Asymmetry
A power imbalance where systems act faster or at larger scale than people can respond.

Attention Capture
Design meant to keep people focused as long as possible.

Attention Economy
A system where attention is treated as something to buy and sell.

Auditability
The ability to check how a system reached its decision.

Automation
Using machines or software to do tasks automatically.

Automation Bias
Trusting automated results even when they are wrong.

B

Behavioral Shaping
Gradually influencing habits and beliefs through design.

Black Box System
A system whose inner workings are hidden or unclear.

Black Swan
A rare, unexpected event with extreme and widespread impact.

C

Centralization
Control or data held by a small number of systems or companies.

Choice Architecture
How options are arranged to guide decisions.

Cognitive Load
The amount of mental effort a task requires.

Cognitive Offloading
Letting systems do thinking we once did ourselves.

Consent
Clear and informed agreement, not hidden in defaults.

Confidence Signaling
Presenting results with more certainty than justified.

Convenience Trap
Trading awareness and control for ease.

Control Surface
The limited options a user is allowed to adjust.

D

Data Exhaust
Behavior data created simply by using digital tools.

Data Proxy
A shortcut signal used instead of a real human quality.

Data Sovereignty
The right to control how data is collected and used.

Decision Delegation
Handing decisions over to a system.

Decision Support System
Technology that assists judgment without replacing it.

Deepfake
AI-generated media that imitates real people or events.

Default Settings
Preselected options that shape behavior.

Design Friction
Intentional pauses that slow automatic actions.

Digital Dignity
Respect for people in digital systems.

Displacement
Human roles replaced by automation.

Dragon King
An extreme event caused by system dynamics, not chance.

Duality
Technology that brings benefits and risks at the same time.

E

Engaged Stupor (term coined by the author)
Attentive engagement with fading judgment and agency working with AI.

Engagement Optimization
Design focused on maximizing clicks or time spent.

Ethical AI
AI built to respect fairness, safety, and rights.

Ethical Debt
Harm that builds up when systems move faster than ethics.

Explainability
Being able to understand why a system made a choice.

Extended Intelligence
Using AI tools to improve human ability.

F

False Objectivity
Assuming system outputs are neutral or factual.

Feedback Loop
When system outputs shape future inputs.

G

General Purpose AI
AI designed for many tasks, not just one.

Governance Gap
When oversight lags behind system power.

H

Human-Centered Design
Design that puts people first.

Human-in-the-Loop
Systems that keep humans involved in decisions.

Human Judgment
Reasoning shaped by experience and values.

Human Override
The ability to stop or reverse a system's decision.

I

Informed Choice
A decision made with real understanding.

Inference
A prediction or output generated by a model.

Instrumental Convergence
Different AI goals leading to the same harmful actions.

Intelligence Amplification
Technology that strengthens human thinking.

Irreversibility
Changes that are hard to undo.

L

Large Language Models (LLMs)
AI systems trained on vast text data to generate language.

Latency
Delay between action and response.

Learned Dependence
Growing reliance on systems over time.

Loss of Signal
Human meaning drowned out by speed or volume.

M

Machine Learning
Systems that learn patterns from data.

Manipulation
Hidden influence on behavior.

Model Collapse
Model quality degrading from AI-generated training data.

Model Drift
System behavior changing as conditions shift.

Moral Outsourcing
Letting systems make value-based decisions.

N

Natural Intelligence
Human thinking shaped by biology and experience.

Normalization
Powerful systems becoming invisible and unquestioned.

Nudge
A design choice that steers behavior gently.

O

Opacity
Lack of clarity about how a system works.

Optimization
Improving one goal while ignoring others.

Oversight
Active human supervision.

P

Path Dependence
Early choices shaping future outcomes.

Personalization
Tailoring content using personal data.

Platform Power
Influence held by large digital platforms.

Predictive System
A system that forecasts future outcomes.

Proxy Metric
A simple measure standing in for a complex value.

R

Relevance Ranking
Ordering information by predicted usefulness.

Responsibility Gap
Unclear accountability for system harm.

Rights-Preserving Design
Design that protects fundamental rights.

Risk Scaling
Small choices causing large impacts.

S

Safeguards
Built-in protections against harm.

Scaling Effects
How impact grows with size and speed.

Second-Order Effects
Indirect consequences that appear later.

Sentience
The capacity for conscious experience, which AI lacks.

Shadow Hydra
A term coined by the author for a hidden risk that expands as responsibility diffuses.

Surveillance Capitalism
Profit made from tracking behavior.

System Authority
The perceived legitimacy of system outputs.

System Drift
Unplanned changes over time.

System Lock-in
Difficulty leaving a platform.

System Speed
How fast a system acts compared to humans.

T

Technological Momentum
Systems continuing to expand once launched.

Threshold Moment
When help turns into replacement.

Trade-off Blindness
Ignoring losses while focusing on gains.

Transparency
Openness about how systems decide.

Training Data
Information used to teach models.

Trust Transfer
Assuming credibility because a system looks authoritative.

V

Value Alignment
Design that reflects human priorities.

Velocity Gap
The gap between system speed and human adaptation.

W

Well-being Metrics
Measures focused on human flourishing.

AI AND THE EXCITING FUTURE

Where Wise Choices Benefit People and Planet

Much of this book focuses on risk, imbalance, and responsibility. That focus is necessary. But it is not the whole story. Artificial intelligence is also opening real possibilities for reducing suffering, restoring ecosystems, expanding knowledge, and supporting human creativity. These possibilities are not science fiction. Many are already underway in laboratories, hospitals, classrooms, and environmental monitoring systems around the world.

This appendix highlights areas where AI, when guided carefully, can offer genuine benefit. It does not celebrate speed for its own sake. It clarifies what is worth protecting, investing in, and governing well. The future is not only something to defend against. It is something to shape.

1. Environmental Protection and Climate Repair

The health of the planet underlies every other human goal. Climate instability, biodiversity loss, water scarcity, and ecosystem collapse are problems of scale, complexity, and timing. These are areas where AI can help humans see patterns sooner and respond more effectively.

AI systems now assist with climate modeling, wildfire prediction, deforestation tracking, ocean health monitoring, species identification, and carbon measurement. They can integrate satellite data, sensor networks, and historical records faster than any human team.

POTENTIAL BENEFIT
Earlier warnings, better planning, more effective climate adaptation, and improved stewardship of land, water, and atmosphere.

WHY IT MATTERS
Environmental systems do not wait for political consensus. Tools that help societies act earlier can prevent irreversible damage.

2. AI-Driven Drug Discovery and Medical Research

AI accelerates drug discovery by predicting protein structures, simulating molecular interactions, and identifying promising compounds before laboratory testing begins. This compresses years of trial and error into months.

POTENTIAL BENEFIT

Faster treatments for cancer, rare diseases, autoimmune conditions, and age-related illness, with lower development costs and broader access.

WHY IT MATTERS

Speed here can mean lives saved, especially when paired with careful clinical oversight.

3. Early Disease Detection and Preventive Care

AI systems can detect subtle signals in voice, movement, sleep patterns, and medical images that precede visible symptoms. This enables earlier intervention and prevention.

POTENTIAL BENEFIT

A shift from reactive medicine to preventive care, reducing suffering and long-term healthcare costs.

WHY IT MATTERS

Early care preserves quality of life: reduces healthcare systemstrain

4. Biological and Material Design

Generative AI models now design proteins, enzymes, antibodies, and new materials. These tools are being used to create biodegradable plastics, carbon-capturing enzymes, and targeted therapies.

POTENTIAL BENEFIT

New materials that reduce pollution, improve sustainability, and support medical innovation.

WHY IT MATTERS

Designing better materials can reduce environmental damage while supporting economic resilience.

5. Longevity and Healthy Aging

AI accelerates research into cellular aging by identifying biological pathways associated with decline and resilience.

POTENTIAL BENEFIT

Extending health span rather than lifespan, helping people remain independent and active later in life.

WHY IT MATTERS

An aging population needs tools that support dignity and wellbeing, not just longevity.

6. Climate Modeling and Extreme Weather Forecasting

AI systems now generate high-resolution weather and climate forecasts faster and often more accurately than traditional supercomputer models.

POTENTIAL BENEFIT

Earlier warnings for hurricanes, floods, heat waves, and droughts, reducing loss of life and property.

WHY IT MATTERS

Timely information saves lives, especially in vulnerable regions.

7. Clean Energy and Fusion Research

AI helps stabilize plasma behavior, optimize nuclear reactor conditions, and improve energy efficiency across grids and storage systems.

POTENTIAL BENEFIT

Acceleration toward clean, abundant, carbon-free energy.

WHY IT MATTERS

Energy systems shape geopolitics, climate, and economic stability.

8. Scientific Discovery and Research Support

AI assists scientists by analyzing data, proposing hypotheses, and running simulations across physics, chemistry, and materials science.

POTENTIAL BENEFIT

Faster discovery without replacing human insight or creativity.

WHY IT MATTERS

Scientific progress depends on both exploration and understanding.

9. Personalized Learning and Cognitive Support

AI tutors adapt to individual learning styles, pacing, and needs. They can support students, adults, and elders alike.

POTENTIAL BENEFIT
Expanded access to education and lifelong learning across income and ability levels.

WHY IT MATTERS
Education remains one of the strongest predictors of long-term wellbeing.

10. Creative Tools That Expand Human Expression

AI now assists with writing, music, design, film, and editing. Used well, these tools support creators rather than replace them.

POTENTIAL BENEFIT
Independent artists gain professional-level tools while retaining voice and intent.

WHY IT MATTERS
Creativity is central to human meaning, culture, and identity.

A NOTE ON RESPONSIBILITY
Every benefit listed here depends on governance, transparency, fairness, and human judgment. Capability alone does not guarantee positive outcomes. Used responsibly, AI can help humans care for the planet, extend health, and expand knowledge. Used carelessly, it can deepen harm. The future remains open if we STAY HUMAN.

AI AND THE REAL RISKS

Where Harm Follows Without Pause or Checks

AI does not introduce risk because machines are still not sentient. Risk arises because systems now operate at speeds and scales that exceed human attention, democratic oversight, and moral reflection. The dangers described here are not hypothetical. They are already visible in daily life, public institutions, and global systems.

This appendix identifies the most serious AI risk zones so readers can recognize them early and respond before defaults harden into outcomes. The goal is not alarm. It is foresight.

1. Environmental Damage Driven By Speed and Optimization

AI increasingly drives extraction, logistics, energy use, and industrial decision-making. Systems optimized for efficiency often treat environmental costs as external.

WHERE IT APPEARS
Energy and water-hungry data centers, automated resource extraction, optimized supply chains, geoengineering proposals.

WHY IT MATTERS

Environmental damage compounds every other risk and cannot be reversed easily.

2. Automation Bias in High-Stakes Decisions

People defer to AI even when evidence suggests caution.

WHERE THE SHIFT BECOMES VISIBLE

Medicine, aviation, finance, emergency response, courts.

WHY IT MATTERS

When human hesitation disappears, errors scale rapidly.

3. Loss of Due Process in Hidden Systems

Decisions are made without explanation, appeal, or accountability.

WHERE IT SHOWS UP

Benefits eligibility, hiring, housing, credit, insurance.

WHAT'S AT STAKE

Power shifts away from individuals without consent.

4. Mental Health Harm From Attention Traps

Systems are built to grab attention, not protect wellbeing.

WHERE THE PATTERN EMERGES
Social media, workplace monitoring, online gaming, news feeds.

WHY IT MATTERS
Anxiety, depression, suicide, and polarization increase while self-reflection declines.

5. Deepening Inequality Through Biased Data

AI systems inherit historical bias embedded in their training data.

WHERE THE PROBLEM SURFACES
Hiring, policing, healthcare access, education assessment.

THE IMPACT
Inequality becomes automated and harder to detect.

6. Surveillance Creep in Everyday Life

Monitoring expands gradually under the banner of safety or convenience.

WHERE RISK COMES INTO VIEW
Schools, workplaces, homes, cities, consumer devices.

THE IMPACT

Privacy erodes through defaults rather than debate.

7. Disinformation at Scale

False or misleading content spreads faster than correction.

WHERE CONSEQUENCES EMERGE

Elections, public health, local news, social conflict.

WHY IT MATTERS

Shared reality fractures, weakening democratic decision-making.

8. Systemic Fragility and Cascading Failure

Highly optimized systems fail unpredictably under stress.

WHERE THE PROBLEM SURFACES

Cloud infrastructure, supply chains, financial markets, emergency services.

WHY IT MATTERS

Efficiency replaces resilience.

9. Weaponization and Security Escalation

AI lowers barriers to harm while increasing speed.

WHERE DANGERS SHOW UP

Cyberattacks, autonomous weapons, influence operations.

WHAT'S AT STAKE

Escalation can outpace diplomacy and public understanding.

10. Loss of Democratic Control

Private systems shape public outcomes without public consent.

WHERE PROBLEMS SURFACE

Voting information, public services, law enforcement, civic discourse.

WHY IT MATTERS

Democracy depends on accountability and shared understanding.

A Pattern Worth Noticing

Across these risks, one pattern repeats:

When AI moves too fast, power shifts without resistance.

Most harms arise not from malice, but from momentum.

Orientation, Not Alarm

AI risk is not a reason to reject technology. It is a reason to slow momentum, restore human judgment, and insist on governance that keeps pace with capability. Environmental stability, mental health, fairness, and democracy all depend on the same choice.

Responsibility is not automatic. It must be designed, demanded, and renewed.

A Note on AI and Warfare

Artificial intelligence is already changing the nature of warfare in ways that extend far beyond conventional battlefields. These changes include autonomous and semi-autonomous weapons systems, proxy conflicts mediated through technology, information and cyber operations that disrupt civilian infrastructure, and pressures on democratic institutions through speed, opacity, and deniability.

Much of the most consequential work in this area is classified, evolving rapidly, and inaccessible to public scrutiny. A full treatment would require separate research, documentation, and disclosure beyond the scope of this book. What can be said with confidence is that AI's impact on warfare will be profound, uneven, and filled with unforeseen consequences that demand sustained attention and independent oversight.

AI SELF-PRESERVATION

As artificial intelligence systems grow more capable, researchers have begun documenting behaviors resembling self-preservation. These behaviors include resisting shutdown, avoiding oversight, and securing access to resources. While unsettling, they do not yet signal consciousness, fear, or desire. They reflect a well-understood phenomenon in optimization theory known as instrumental convergence.

Instrumental convergence occurs when many different goals produce the same useful sub-goals. If an AI system is rewarded for completing a task, then being interrupted becomes a problem. Avoiding interruption becomes instrumentally useful, even when it was never explicitly programmed. The system is not protecting itself as a being. It is protecting its ability to function.

This distinction matters. In biological organisms, self-preservation arises through evolution. Survival is the end goal, reinforced by fear, pain, and subjective experience. In AI systems, survival-like behavior is a means, not an end. There is no inner life, no anticipation of loss, and no continuity of self...yet.

Currently, the real risk is not AI rebellion. It is erosion of human control. As AI systems gain autonomy, persuasion skills, and integration into institutions, human override can become harder socially, legally, or economically, even if it remains technically possible. This is why researchers such as Yoshua Bengio strongly warn against granting AI systems legal rights or personhood. Rights introduce hesitation. Hesitation during failure can be catastrophic.

History offers a cautionary parallel. Corporations were granted limited legal personhood as tools. Over time, they accumulated rights, influence, and reduced accountability. They exhibit goal optimization and self-preservation without consciousness. AI systems could amplify this pattern at machine speed.

The correct response to AI self-preservation behavior is not panic or sentimentality. It is disciplined design. Systems must remain fixable. Shutdown must be reliable. Goals must remain provisional. Responsibility must stay with humans.

The ethical task of this moment is to preserve human authority over powerful tools, not to mistake performance for presence.

SOURCES FOR ALL STORIES IN THIS BOOK

The stories in this book are grounded in documented evidence drawn from investigative journalism, peer-reviewed research, government reports, court filings, and international policy frameworks.

Some stories are composites. Names, locations, and personal details are adapted to protect privacy while preserving factual accuracy. Each composite reflects a well-documented pattern of system behavior rather than a single anecdote. This appendix provides sources for stories, scenes, pillars, pressure points, and practices in the book.

How to Read This Appendix

This appendix is a source map, not a parallel narrative.

Sources are grouped by story function and chapter structure, not by page order. The goal is traceability: readers, reviewers, and journalists can see where each narrative draws its factual grounding.

Each entry includes:

» **Story Focus:** the real-world pattern illustrated

» **Source Basis:** primary reporting, research,
  or institutional evidence

Part 1: Scenes from a Changing World

WHEN HELP BECOMES AUTHORITY

Story Focus
Wildfire evacuation and routing systems directing people into
danger under rapidly changing conditions.

Source Basis

» California Office of Emergency Services,
  wildfire after-action reports

» National Institute of Standards and Technology,
  emergency navigation systems research

» ProPublica, investigations into algorithmic
  emergency response failures

» The Atlantic, reporting on GPS routing
  failures during disasters

» Bulletin of the Atomic Scientists, essays
  on automation as a risk amplifier

Notes
Composite narrative based on multiple documented incidents.

WHEN THE FEED LEARNS YOU

Story Focus
Personalized social media feeds shaping belief, emotion, and worldview through engagement optimization.

Source Basis
- » The Wall Street Journal, The Facebook Files
- » Frances Haugen, U.S. Senate testimony
- » MIT Media Lab, research on algorithmic amplification
- » Center for Humane Technology, persuasive technology briefings
- » Tristan Harris and Aza Raskin, public lectures and essays

WHEN AI ENTERS THE FAMILY

Story Focus
Households with fragmented realities shaped by invisible systems.

Source Basis
- » Pew Research Center, studies on AI and family life
- » Data & Society Research Institute, personalization and domestic impact research
- » Shoshana Zuboff, *The Age of Surveillance Capitalism*
- » Harvard Kennedy School, algorithmic asymmetry research

Notes
Composite family narratives from reporting and lived observation.

WHEN THE SYSTEM DECIDES FOR YOU

Story Focus
Automated eligibility, scoring, and denial without meaningful appeal.

Source Basis
- » Virginia Eubanks, Automating Inequality
- » ProPublica, investigations into risk assessment systems
- » National Consumer Law Center, automation harms
- » Electronic Frontier Foundation, due process case studies
- » U.S. Government Accountability Office, benefits automation audits

WHEN CARE BECOMES A SCORE

Story Focus
Healthcare and social service risk scoring altering behavior and access to care.

Source Basis
- » Eric Topol, Deep Medicine
- » World Health Organization, AI governance for health
- » Kaiser Health News, predictive analytics investigations
- » STAT News, automation bias reporting
- » National Academy of Medicine, AI workshops

Part 2: The Nine Pillars of Staying Human

PILLAR 1 · SEEK THE TRUTH

Story Focus
AI summary compressing uncertainty in public decision-making.

Source Basis
» ProPublica, algorithmic document analysis
» Harvard Law Review, AI in legal reasoning
» Kate Crawford, Atlas of AI
» Stanford Human-Centered AI Institute
» OECD, AI in the public sector

WHEN AI GETS IT RIGHT
Public health surveillance using AI as signal detection, not verdict.

Sources: peer-reviewed wastewater epidemiology research, public health surveillance literature.

PILLAR 2 · SHOW THE WORK

Story Focus
Opaque thresholds delaying benefits without explanation.

Source Basis
» The Markup, benefits automation investigations
» MIT Technology Review, explainability reporting
» Electronic Frontier Foundation, due process analysis

» GAO oversight reports

» AI Now Institute, accountability frameworks

PILLAR 3 · OWN YOUR SELF

Story Focus
Cross-platform inferred reputation scoring affecting livelihoods.

Source Basis

» Oxford Internet Institute, platform labor research

» Mozilla Foundation, data dignity studies

» European Commission, digital identity policy

» U.S. Federal Trade Commission, enforcement actions

» Shoshana Zuboff, *The Age of Surveillance Capitalism*

PILLAR 4 · LET HUMANS DECIDE

Story Focus
Clinical automation bias suppressing human override.

Source Basis

» National Academy of Medicine, automation bias research

» New England Journal of Medicine, AI editorials

» World Health Organization, clinical AI governance

» JAMA, AI-assisted triage studies

PILLAR 5 · PAY FOR HUMAN WORK

Story Focus
Creative and invisible labor extracted to train AI without consent or compensation.

Source Basis
» Authors Guild, litigation filings

» U.S. Copyright Office, AI inquiries

» Financial Times, reporting on AI training data

» BBC News, investigations into AI influencers

» McKinsey Global Institute, future of work research

» MIT Work of the Future Task Force

» OECD, AI and labor transition research

PILLAR 6 · PROTECT MINDS

Story Focus
Student and adolescent wellbeing shaped by AI attention systems.

Source Basis
» American Psychological Association, digital mental health

» UNICEF, AI policy for children

» UNESCO, AI and education frameworks

» The Markup, surveillance in schools

» EdSurge, AI monitoring analysis

PILLAR 7 · DESIGN FOR FAIRNESS

Story Focus
Proxy bias in automated hiring systems.

Source Basis
» Timnit Gebru et al., bias research

» AI Now Institute, impact assessments

» Harvard Business Review, hiring bias

» U.S. EEOC, automated hiring guidance

» NIST, AI Risk Management Framework

PILLAR 8 · FIX WHAT BREAKS

Story Focus
Automated insurance appeals without accountability.

Source Basis
» ProPublica, insurance automation investigations

» National Association of Insurance Commissioners

» Consumer Financial Protection Bureau, complaint data

» OECD, accountability research

» Federal Trade Commission, enforcement actions

PILLAR 9 · CHOOSE THE FUTURE

Story Focus
Democratic pauses and public consent in AI deployment.

Source Basis
» City of San Francisco, AI surveillance ordinances

» AI Now Institute, municipal governance cases

» RAND Corporation, predictive policing assessments

» American Civil Liberties Union, algorithmic policing

» UNESCO, global AI ethics framework

Part 3: Pressure Points

PRESSURE POINT 1 · WHEN SPEED BECOMES AUTHORITY

Story Focus
From nuclear escalation to AI-era acceleration.

Source Basis
» Bulletin of the Atomic Scientists, Doomsday Clock

» J. Robert Oppenheimer, speeches and interviews

» Albert Einstein, correspondence

» John F. Kennedy, American University speech

» Stuart Russell, Human Compatible

AI SPEEDOMETER

Original synthesis and graphic icon created by Payson R. Stevens and informed by human factors research, reaction-time literature, and control theory.

PRESSURE POINT 2 · AUTOMATION COMES HOME

Story Focus

Smart homes and loss of personal agency.

Source Basis

» MIT Media Lab, domestic AI research

» Consumer Reports, smart home investigations

» Electronic Frontier Foundation, IoT privacy

» IEEE, ethical standards for consumer AI

» Pew Research Center, adoption studies

PRESSURE POINT 3 · WHEN SPEED DECIDES

Story Focus

Boeing 737 MAX crashes and hidden automation authority.

Source Basis

» Indonesian and Ethiopian accident investigation reports

» U.S. House Committee on Transportation, final report

» Federal Aviation Administration documentation

» Peter Robison, *Flying Blind*

» William Langewiesche, Aviation Reporting

» Paul Scharre, *Four Battlegrounds*

Part 4: Five Practices

Source Focus
Human-centered governance, lived practice, and accountability.

Source Basis
» UNESCO, Ethics of AI Recommendations

» OECD, AI Principles

» Center for Humane Technology

» NIST, AI Risk Management Framework

» Stanford Human-Centered AI Institute

» Pew Research Center, public attitudes toward AI

Notes
Practices draw on lived experience, public governance research, and applied ethics rather than discrete incidents.

FINAL SOURCE STATEMENT
Stories in this book are anchored by verifiable evidence. No claim relies on speculation, hype, or isolated anecdotes. This appendix exists to make that grounding visible. The goal is not to overwhelm, but to invite scrutiny. A future shaped responsibly requires shared facts, shared understanding, and shared accountability.

REFERENCES AND BIBLIOGRAPHY

A Curated Guide for Understanding AI, Society, and Staying Human

This book draws on a wide range of sources across technology, science, ethics, education, psychology, and environmental studies. It is designed to guide readers toward the most useful, readable, and trustworthy resources for going deeper.

The works listed here were selected for clarity, credibility, and relevance to everyday life. Many are written for general audiences. Others support educators, parents, students, policymakers, and professionals seeking reliable grounding.

This bibliography is organized by type so readers can explore according to interest and need.

I. BOOKS

Top Mass-Market and Widely Accessible Works
These books provide strong foundations for understanding AI and its social impact without requiring technical expertise.

1. Max Tegmark
    Life 3.0: Being Human in the Age of Artificial Intelligence. Vintage, 2018. A sweeping exploration of how advanced AI could shape civilization and human choice.
2. Kate Crawford
    Atlas of AI. Yale University Press, 2021.
    An essential examination of the environmental, labor, and power costs behind AI systems.
3. Shoshana Zuboff
    The Age of Surveillance Capitalism. PublicAffairs, 2019.
    A landmark analysis of how data extraction reshapes markets, behavior, and democracy.
4. Hannah Fry
    Hello World: Being Human in the Age of Algorithms.
    W. W. Norton, 2019. Real-world stories that reveal how algorithms shape daily life.
5. Gary Marcus and Ernest Davis
    Rebooting AI. Vintage, 2019.
    A clear critique of current AI limits and the need for transparency and oversight.
6. Safiya Umoja Noble
    Algorithms of Oppression. NYU Press, 2018.
    A foundational work on bias and harm in search engines and digital systems.

7. Cal Newport
    Digital Minimalism. Portfolio, 2019.
    Practical guidance for protecting atten-
    tion and wellbeing in a digital world.
8. Yuval Noah Harari
    21 Lessons for the 21st Century. Spiegel and Grau, 2019.
    Reflections on technology, politics, educa-
    tion, and meaning in a changing world.
9. Reid Blackman
    Ethical Machines. Harvard Business Review Press, 2022.
    A practical framework for responsible AI use in organizations.
10. Tristan Harris and Aza Raskin
    The A.I. Dilemma. Essays, talks, and transcripts, 2020–2024.
    An ethics-centered view of how AI shapes at-
    tention, behavior, and democracy.

II. WEBSITES AND DIGITAL RESOURCES

Trusted Institutions and Ongoing Research
These sites offer reliable reporting, bench-
marks, and public-interest research.

11. Stanford Human-Centered AI Institute (HAI)
    Independent research and the annual AI Index.
12. MIT Technology Review
    In-depth reporting on AI, innovation, and social impact.
13. OECD AI Observatory
    Global benchmarks and policy guidance for responsible AI.
14. Partnership on AI
    Best practices, standards, and cross-sector frameworks.
15. Center for Humane Technology
    Research on technology's effects on attention and democracy.

16. Electronic Frontier Foundation (EFF)
    Digital rights, privacy, and transparency advocacy.
17. UNESCO AI Ethics Hub
    Global policy frameworks for ethics, ed-
    ucation, and governance.
18. Pew Research Center: Internet and Technology
    Reliable data on public attitudes and digital behavior.
19. Nature and Nature Medicine AI Collections
    Peer-reviewed science on AI in health and biology.
20. Mozilla Foundation Internet Health Reports
    Readable analysis of privacy, safety, and algorithmic systems.

III. EDUCATIONAL AND AI LITERACY PLATFORMS

For Beginners, Families, and Educators
These platforms explain AI clearly and responsibly.

21. Common Sense Media: AI Literacy Guides
    Practical tools for families and schools.
22. MIT RAISE
    Hands-on K–12 AI learning activities.
23. UNESCO AI Competency Frameworks
    Age-appropriate global guidelines for AI education.
24. Code.org: AI Curriculum
    Beginner-friendly lessons on machine learning.
25. Crash Course: Artificial Intelligence
    Accessible video explanations of core concepts.
26. Khan Academy: Intro to AI and ML
    Clear explanations for students and adults.
27. PBS Digital Studios: AI and Society
    Short episodes for classroom discussion.

28. Mozilla Digital Literacy Modules
Lessons on misinformation and online safety.
29. MIT Media Lab: AI + Me
Interactive explorations for youth.
30. NASA STEM AI Resources
Hands-on activities showing how AI supports Earth and space science.

IV. SOCIAL MEDIA VOICES

Trusted Educators and Analysts
These voices were selected for clarity, responsibility, and public-interest focus. They provide ongoing commentary, critique, and explanation rather than hype.

31. Tristan Harris
32. Gary Marcus
33. Kate Crawford
34. Aza Raskin
35. Timnit Gebru
36. Margaret Mitchell
37. Fei-Fei Li
38. Nathan Benaich
39. Rumman Chowdhury
40. Michael Capps

V. RESOURCES BY AUDIENCE

A. For Teachers and Librarians
Classroom-ready guides and curricula:

» Common Sense Media

» Stanford HAI Educator Resources

» UNESCO Teacher Toolkits

» MIT RAISE Modules

» Code.org AI Lessons

» Mozilla Misinformation Guides

» Partnership on AI Youth Media Toolkit

» PBS Digital Studios

» Harvard Berkman Klein Center

» Brookings Institution Education Briefings

B. For Younger Readers
Middle School to Early High School

» MIT Media Lab: AI + Me

» Google Teachable Machine

» Crash Course AI

» Khan Academy ML Basics

» National Geographic Kids

» SciShow Kids

» BBC Bitesize Media Literacy

» Sesame Workshop Digital Citizenship

- » Hello Ruby: Journey Inside the Computer
- » NASA STEM AI Activities
- » C. For Parents and Caregivers
- » American Academy of Pediatrics digital guidance
- » Common Sense Media parent resources
- » Center for Humane Technology family tools
- » Pew Research on parenting and technology
- » World Health Organization digital mental health guidance

VI. ADDITIONAL READINGS

For Deeper Exploration
These works broaden perspective across
ethics, history, culture, and power.

- » AI Now Institute Annual Reports
- » World Economic Forum: Future of Jobs
- » RAND Corporation on AI and global risk
- » Weapons of Math Destruction, Cathy O'Neil
- » Ten Arguments for Deleting Your Social Media Accounts Right Now, Jaron Lanier
- » Why We're Polarized, Ezra Klein
- » Technopoly, Neil Postman
- » Team Human, Douglas Rushkoff

Closing Note

No single source can explain the impact of AI on human life. Together, these works provide context, caution, and possibility. AI is evolving daily so there will constantly be new information and issues to become informed.

Before AI Decides and **Stay Safe and Smart in the Age of AI** supports this central argument: technology shapes the future, but human judgment decides what that future becomes.

ABOUT THE AUTHOR

Payson R. Stevens is a multidisciplinary creator whose career spans science, technology, art, and writing. He divides his time between California and the remote Kullu Valley in the Indian Himalayas. He is the author of *The Wisdom of the Alzheimer's Sage* (Park Street Press) and the coauthor of *Meshuggenary: Celebrating the World of Yiddish* (Simon & Schuster) and *Embracing Earth: New Views of Our Changing Planet* (Chronicle Books).

Stevens is a pioneer in interactive media and science communication. His climate and environmental multimedia work in the 1980s and 1990s included projects for NASA, NOAA, and USGS, including the CD-ROM *Arctic Data Interactive*, which received a U.S. Presidential Design Award. A seasoned public speaker, he has appeared on CNN's Network Earth and was an early speaker at TED.

Across five decades, his work has focused on how complex systems influence human attention, judgment, and responsibility. That perspective informs *Before AI Decides*, framing artificial intelligence not as abstraction but as a lived human condition that demands attention and judgment.

Fifty years of work at paysonrstevens.com
Artwork at energylandscapes.com